Lecture Notes in Elementary Real Analysis

ROHAN DALPATADU

Order this book online at www.trafford.com
or email orders@trafford.com

Most Trafford titles are also available at major online book retailers.

© Copyright 2015 Rohan Dalpatadu.
All rights reserved. No part of this publication may be reproduced, stored in a retrieval system, or transmitted, in any form or by any means, electronic, mechanical, photocopying, recording, or otherwise, without the written prior permission of the author.

Print information available on the last page.

ISBN: 978-1-4907-6472-6 (sc)
ISBN: 978-1-4907-6471-9 (e)

Library of Congress Control Number: 2015915458

Because of the dynamic nature of the Internet, any web addresses or links contained in this book may have changed since publication and may no longer be valid. The views expressed in this work are solely those of the author and do not necessarily reflect the views of the publisher, and the publisher hereby disclaims any responsibility for them.

Any people depicted in stock imagery provided by Thinkstock are models,
and such images are being used for illustrative purposes only.
Certain stock imagery © Thinkstock.

Trafford rev. 10/09/2015

www.trafford.com

North America & international
toll-free: 1 888 232 4444 (USA & Canada)
fax: 812 355 4082

CONTENTS

PREFACE vii

CHAPTER 1 The Real Number System 1
1.1 Sets and Operations on Sets 1
1.2 Relations and Functions 3
1.3 Mathematical Induction 5
1.4 The Real Number System 8
1.5 Countable and Uncountable Sets 18

CHAPTER 2 Sequences and Series of Real Numbers 27
2.1 Convergent Sequences 27
2.2 Limit Theorems 35
2.3 Monotonic Sequences 45
2.4 Subsequences and the Bolzano-Weierstrass Theorem 50
2.5 Limit Superior and Limit Inferior 56
2.6 Cauchy Sequences 65
2.7 Series of Real Constants 70
2.8 Further Tests for Convergence and Absolute Convergence 78

CHAPTER 3 Limits and Continuity 90
3.1 Structure of Point Sets in $\mathbb{R}$ 90
3.2 Limit of a Function 102
3.3 Continuous Functions 114
3.4 Uniform Continuity 122
3.5 Discontinuities and Monotonic Functions 125

CHAPTER 4 Differentiation 133
4.1 The Derivative 133
4.2 Mean Value Theorems 139

CHAPTER 5 Integration 151
5.1 The Riemann Integral 151
5.2 Properties of the Riemann Integral 161
5.3 Fundamental Theorem of Calculus 168
5.4 Improper Riemann Integrals 176

CHAPTER 6 Sequences and Series of Functions 186
6.1 Sequence of Functions 186
6.2 Series of Functions 201
6.3 Power Series and Special Functions 213

BIBLIOGRAPHY 235

PREFACE

These lecture notes are the ones presented to my Elementary Real Analysis classes in the fall semester of 2014 and the spring semester of 2015 and are solely based on two well written texts on the subject: Principles of Mathematical Analysis by Walter Rudin, 3rd Edition, McGraw-Hill, Inc. 1976 and Introduction to Real Analysis by Manfred Stoll, Addison Wesley Longman, Inc. 2001. The majority of examples and assignments were also from these two texts.

The first three chapters were covered in the fall semester and the last three chapters in the spring semester. Due to the expanded nature of the notes, the material covered in the two semesters appear to be somewhat less than that of typical courses across the nation. However, this enabled most of the students to obtain a deeper understanding of the subject and also the techniques used in the proofs.

I have only assumed that the students have been introduced to the rational numbers and their properties and developed the real number system based on this understanding. The assignments and the examples used just the functions that have been introduced in the course, e.g., the exponential function was only used in the examples after it had been defined along with its properties. The trigonometric functions and hyperbolic functions were defined almost towards the end of the course. However, anyone wishing to use these notes can always include problems and examples based on functions to be introduced later on.

Chapter 1 introduces the Real Number System with a Completenes Axiom; I have deliberately avoided Dedekind cuts to make the completeness more understandable to undergraduate students. Furthermore, the Principle of Mathematical Induction was also presented because of its varied applications. Chapter 2 concentrates on the development of sequence and series of constants and explains in detail almost all of the essential results on them. Chapter 3 is on limits and continuity of real valued functions and I have briefly introduced Point Set Topology in order to present elegant proofs of some of the theorems. Chapter 4 discusse the basic ideas of differentiation with some Mean Value Theorems. In Chapter 5 on integration, I have deliberately avoided the Riemann Stieltjes Integral because the results and proofs are quite similar and also because this concept can be handled quite easily, if the students have a solid understanding of the basic ideas. Chapter 6 is where, the most important sections of Elementary Real Analysis are covered, i.e., Sequences and Series of Functions. Here, we are able to introduce the Transcendental Functions with their properties. However, due to the expanded nature of the course, I was not able to proceed any further than this.

The lecture notes provided here should give most students a solid background in Elementary Real Analysis in order for them to be able to complete a course in (graduate level) Real Analysis. As a last note I would like to add that the text by Walter Rudin is ideal for an introductory course in (real) analysis for most advanced undergraduate mathematics majors, whereas Manfred Stoll's text would be quite useful to almost all students. I have tried to go one step further by providing more details and examples and presenting the material without assuming results to be presented later on.

CHAPTER

The Real Number System

1.1 Sets and Operations on Sets

In this section, we shall define sets and operations on sets that the reader is already familiar with and will use standard notation. Examples are deemed not necessary for this section and have not been provided.

1.1D1 Definition: Sets

A **set** is a well defined collection of objects and can be described by listing its objects or elements and also may be defined as the collection of all elements x satisfying a given property. Thus the notation

$$A = \{x \mid P(x)\}$$

defines A to be the set of all elements x satisfying the condition $P(x)$. This reads as

"A equals the set of all elements x such that $P(x)$."

Notation: $x \in A$ means that x is an **element** of the set A. $x \notin A$ means that x is **not an element** of the set A.

We may also use the notation $A = \{x \in X \mid P(x)\}$ to mean that only the elements x from the set X are being used.

We will assume that the set of **natural numbers** or **positive integers** have already being defined. This set will be denoted by $\mathbb{N}$ or $\{1, 2, 3, ...\}$.

We will also assume that the set of **integers** have already being defined. This set will be denoted by $\mathbb{Z}$ or $\{0, \pm 1, \pm 2, \pm 3, ...\}$.

Furthermore, the set of rational numbers with the binary operations of addition and multiplication will be denoted by $\mathbb{Q}$. We shall also assume all the properties of $\mathbb{Q}$.

The set without any members is called the **empty (or null) set** and is denoted by ϕ.

1.1D2 Definition: Equality of Sets

Two sets A and B are said to be **equal** if every element of A is an element of B and every element of B is an element of A. In this case, we write $A = B$. $A \neq B$ means that the set A is **not equal** to the set B.

1.1D3 Definition: Subsets of Sets

A set A is said to be a **subset** of a set B or A is said to be contained in B if every element of A is also an element of B. In this case, we use the notation: $A \subseteq B$. If $A \subseteq B$ and $A \neq B$, then we say that A is a **proper subset** of B and use the notation $A \subset B$. Note that the empty set ϕ is a subset of any set A.

Note: $A = B$ if and only if $A \subseteq B$ and $B \subseteq A$.

1.1D4 Definition: Union, Intersection, and Complement of Sets

The **union** of sets A and B is the set of all elements that belong to A or B or both A and B and is denoted by $A \cup B$. Thus,

$$A \cup B = \{x \mid x \in A \text{ or } x \in B\}.$$

The **intersection** of sets A and B is the set of all elements that belong to both A and B and is denoted by $A \cap B$. Thus,

$$A \cap B = \{x \mid x \in A \text{ and } x \in B\}.$$

The **relative complement** of A in B or the **complement** of A **relative to** B, denoted by $B \setminus A$ is the set of all elements in B that are not elements of A. Thus

$$B \setminus A = \{x \in B \mid x \notin A\}.$$

If A is a subset of some fixed set X consisting of all the elements under consideration, then $X \setminus A$ is referred to as the **complement** of A and is denoted by A^c. Thus,

$$A^c = \{x \mid x \notin A\}.$$

1.1T1 Theorem: Distributive and DeMorgan's Laws

Let A, B, and C be sets. Then

(a) $A \cap (B \cup C) = (A \cap B) \cup (A \cap C)$.

(b) $A \cup (B \cap C) = (A \cup B) \cap (A \cup C)$.

(c) $C \setminus (A \cup B) = (C \setminus A) \cap (C \setminus B)$.

(d) $C \setminus (A \cap B) = (C \setminus A) \cup (C \setminus B)$.

(a) and (b) are referred to as the **distributive laws**, whereas (c) and (d) are known as **DeMorgan Laws**.

Proof: We will provide the proofs of (a) and (c). The proofs of (b) and (d) are left as an exercise.

(a) $x \in A \cap (B \cup C) \Leftrightarrow x \in A$ and $x \in B \cup C$

$\Leftrightarrow x \in A$ and

$\Leftrightarrow (x \in A$ and $x \in B)$ or $(x \in A$ and $x \in C)$

$\Leftrightarrow x \in (A \cap B) \cup (A \cap C).$

This completes the proof of (a).

(c) $x \in C \setminus (A \cup B) \Leftrightarrow x \in C$ and $x \notin (A \cup B)$

$\Leftrightarrow x \in C$ and $(x \notin A$ and $x \notin B)$

$\Leftrightarrow (x \in C$ and $x \notin A)$ and $(x \in C$ and $x \notin B)$

$\Leftrightarrow x \in C \setminus A$ and $x \in C \setminus B$

$\Leftrightarrow x \in (C \setminus A) \cap (C \setminus B).$

This completes the proof of (c).

1.1D5 Definition: The Power Set

Let A be a set. The set consisting of all the subsets of A is called the **power set** of A and is denoted by $\mathrm{P}(A)$. If a set has n elements, then its power set consists of 2^n elements, i.e., the set has 2^n subsets.

1.1D6 Definition: Cartesian Product

Let A and B be sets. Then the **Cartesian product** of A and B, denoted by $A \times B$, is the set of all ordered pairs (a,b), where the first component a is an element of A and the second component b is an element of B. Thus,

$$A \times B = \{(a,b) \mid a \in A \text{ and } b \in B\}.$$

1.2 Relations and Functions

1.2D1 Definition: Relations and Functions

Let A and B be sets. Then a subset of the Cartesian product $A \times B$ is called a **relation** from A into B. If ρ is a relation from A into B, then the **domain** of ρ, denoted by $Dom(\rho)$ is given by

$$Dom(\rho) = \{a \mid (a,b) \in \rho\}.$$

The **range** of ρ, denoted by $Ran(\rho)$, is given by

$$Ran(\rho) = \{b \mid (a,b) \in \rho\}.$$

A subset F of the Cartesian product $A \times B$ is called a **function** from A into B if the following condition is also satisfied:

$$(x, y_1) \in F \text{ and } (x, y_2) \in F \Rightarrow y_1 = y_2.$$

Notation: The domain and the range of F have already been defined because a function is also a relation. If F is a function and $(x, y) \in F$, then we may use the notation $y = f(x)$, and may refer to the function as f or $y = f(x)$ or simply $f(x)$. If f is a function from A to (into) B, we say that f maps A to (into) B and use the notation: $f : A \to B$.

1.2D2 Definition: One-to-one and Onto Functions

Let $f : A \to B$. The function f is said to be **onto** if $Ran(f) = B$. In this case, we say that f is a function **(mapping)** from A onto B. The function f is said to be **one-to-one** $(1-1)$ if the following condition is satisfied:

$$(x_1, y) \in f \text{ and } (x_2, y) \in f \Rightarrow x_1 = x_2.$$

Note: The function f is one-to-one if $x_1 \neq x_2 \Rightarrow f(x_1) \neq f(x_2)$.

1.2D3 Definition: Image and Inverse Image

Let $f : A \to B$, $E \subseteq A$ and $H \subseteq B$.

(a) The **image** of E under f, denoted $f(E)$, is given by

(b) The **inverse image** of H under f, denoted $f^{-1}(H)$, is given by

$$f^{-1}(H) = \{x \in A \mid f(x) \in H\}.$$

1.2D4 Definition: The Inverse Function

Let $f : A \to B$ be $1-1$. Then the **inverse function** of f, denoted by f^{-1}, is the function: $f^{-1} = \{(y, x) \in B \times A \mid f(x) = y\}$.

The **domain** of f^{-1} is the range of f and the **range** of f^{-1} is the domain of f.

Note: $x = f^{-1}(y) \Leftrightarrow y = f(x)$.

1.2D5 Definition: Composition of Functions

Let $f : A \to B$ and $g : B \to C$. Then the **composition** of g with f, denoted $g \circ f$, is the function from A into C defined by

$$g \circ f = \{(x, z) \mid z = g(y) \text{ where } y = f(x),\ x \in A\}$$

The **domain** and **range** of $g \circ f$ are given by:

$$Dom(g \circ f) = \{x \in A \mid f(x) \in B\}$$

and

$$Ran(g \circ f) = \{z \in C \mid z = g(f(x)),\ x \in A,\ f(x) \in B\}.$$

1.2D6 Definition: The Identity Function

The **identity function** on a set *A*, is the function $i_A : A \to A$ given by

$$i_A = \{(x, x) \mid x \in A\}.$$

This function is clearly one-to-one and onto.

1.2T1 Theorem:

$f : A \to B$ be $1-1$. Then

(a) $f^{-1} \circ f = i_A$.

(b) $f \circ f^{-1} = i_{Ran(f)}$.

Proof:

(a) $x \in A,\ y = f(x) \Rightarrow (f^{-1} \circ f)(x) = f^{-1}(f(x)) = f^{-1}(y) = x.$

(b) Similar to (a).

1.2A1 Assignment:

(a) Let $f : A \to B$ and let $C \subseteq A$. Prove that $f(A) \setminus f(C) \subseteq f(A \setminus C)$ and provide an example where $f(A) \setminus f(C) \neq f(A \setminus C)$.

(b) Let $f : A \to B$ and $g : B \to A$ satisfying: $g \circ f = i_A$. Show that *f* is one-to-one. Should *f* be onto? Justify!

(c) $f : A \to B$ and $g : B \to C$ are one-to-one functions. Prove that $(g \circ f)^{-1} = f^{-1} \circ g^{-1}$ on $Ran(g \circ f)$.

(d) $f : A \to B$ and $g : B \to C$ are onto functions. Show that $g \circ f$ is onto.

1.3 Mathematical Induction

Mathematical induction is a useful tool in proving a statement, identity, or inequality involving the positive integer *n*. We will provide two versions of mathematical induction and the well-ordering principle in this section.

Note: The Well-Ordering Principle

Every nonempty subset of $\mathbb{N}$ has a smallest element. This statement is taken as an axiom for the natural numbers and may be restated as follows:

$$\phi \neq A \subseteq \mathbb{N} \Rightarrow \text{ there exists } n \in A \text{ such that } n \leq k \text{ for all } k \in A.$$

1.3T1 Theorem: The Principle of Mathematical Induction

Let $P(n)$ be a statement about n for each $n \in \mathbb{N}$. Suppose

(a) $P(1)$ is true,

(b) $P(k+1)$ is true whenever $P(k)$ is true, where $k \in \mathbb{N}$.

Then $P(n)$ is true for all $n \in \mathbb{N}$.

Proof: Suppose the conclusion is false, i.e., there exists $n \in \mathbb{N}$ such that $P(n)$ is false. Let

$$A = \{k \in \mathbb{N} \mid P(k) \text{ is false}\}.$$

The set A is a nonempty subset of the natural numbers and must have a smallest element, say m. Since $P(1)$ is true, $m > 1$. Since m is the smallest element of A, $P(m-1)$ is true. But then (b) implies that $P(m)$ is true. This is a contradiction. Therefore, $P(n)$ is true for all $n \in \mathbb{N}$.

1.3E1 Example:

Prove that: $1^3 + 2^3 + \ldots + n^3 = \dfrac{n^2(n+1)^2}{4}$ for all $n \in \mathbb{N}$.

Let $P(n)$: $1^3 + 2^3 + \ldots + n^3 = \dfrac{n^2(n+1)^2}{4}$, $n \in \mathbb{N}$. Then $P(1)$: $1^3 = \dfrac{(1^2)(2^2)}{2^2}$, is true. Suppose $P(k)$ is true, where $k \in \mathbb{N}$. Then

$$\begin{aligned} 1^3 + 2^3 + \ldots + k^3 + (k+1)^3 &= \frac{k^2(k+1)^2}{4} + (k+1)^3 \\ &= \frac{(k+1)^2}{4}\left(k^2 + 4k + 4\right) \\ &= \frac{(k+1)^2((k+1)+1)^2}{4}. \end{aligned}$$

Thus, $P(k+1)$ is true. Therefore, by the principle of mathematical induction, $P(n)$ is true for all $n \in \mathbb{N}$.

Note: If we change the hypothesis (a) in **1.3T1** to $P(m)$ is true, where $m \in \mathbb{Z}$, then we have the **Modified Principle of Mathematical Induction**, stated as follows:

Let $P(n)$ be a statement about *n* for each $n \in \mathbb{Z}$. Suppose

(a′) $P(m)$ is true, where $m \in \mathbb{Z}$,

(b′) $P(k+1)$ is true whenever $P(k)$ is true, where $k \in \mathbb{Z},\ k \geq m$.

Then $P(n)$ is true for all $n \in \mathbb{Z},\ n \geq m$.

This is easily proved by defining a new statement: $Q(n) = P(n+m-1),\ n \in \mathbb{N}$.

1.3E2 Example:

Prove that: $1^3 + 2^3 + \ldots + n^3 < \frac{1}{2}n^4$ for all $n \in \mathbb{N},\ n \geq 3$.

Let $P(n)$: $1^3 + 2^3 + \ldots + n^3 < \frac{1}{2}n^4$ for all $n \in \mathbb{N},\ n \geq 3$.

Then $P(3)$: $1^3 + 2^3 + 3^3 < \frac{1}{2}3^4$, is true. Suppose $P(k)$ is true, where $k \in \mathbb{Z},\ k \geq 3$. First we show that $2k^3 > 2k+1$ if $k \in \mathbb{Z},\ k > 1$. This follows easily because,

$$2k^3 - 2k - 1 = 2k(k^2-1) - 1 > 2k - 1 > 2 - 1 > 0 \text{ if } k > 1.$$

Therefore,

$$\begin{aligned} 1^3 + 2^3 + \ldots + k^3 + (k+1)^3 &< \frac{1}{2}k^4 + (k+1)^3 \\ &= \frac{1}{2}\left(k^4 + 2k^3 + 6k^2 + 6k + 2\right) \\ &< \frac{1}{2}\left(k^4 + 4k^3 + 6k^2 + 4k + 1\right) < \frac{1}{2}(k+1)^4. \end{aligned}$$

Thus, $P(k+1)$ is true. Therefore, by the modified principle of mathematical induction, $P(n)$ is true for all $n \in \mathbb{N},\ n \geq 3$.

1.3T2 Theorem: The Second Principle of Mathematical Induction

Let $P(n)$ be a statement about *n* for each $n \in \mathbb{N}$. Suppose

(a) $P(1)$ is true,

(b) $P(k+1)$ is true whenever $P(1),\ P(2), \ldots,\ P(k)$ is true, where $k \in \mathbb{N}$.

Then $P(n)$ is true for all $n \in \mathbb{N}$.

Proof: Exercise.

1.3E3 Example: Let $f:\mathbb{N}\to\mathbb{N}$ be defined by $f(1)=1$, $f(2)=2$, and

$$f(n+2)=\frac{f(n+1)+f(n)}{2} \quad \text{for all } n\in\mathbb{N}.$$

Prove that: $1\le f(n)\le 2$ for all $n\in\mathbb{N}$.

Let $P(n)$: $1\le f(n)\le 2$ for all $n\in\mathbb{N}$. Then $P(1)$: $1\le f(1)\le 2$ and $P(2)$: $1\le f(2)\le 2$ are true. Suppose for $k\ge 1$, $P(1)$, $P(2)$, ..., $P(k)$, $P(k+1)$ are true. Then

$$1=\frac{1+1}{2}\le\frac{f(k+1)+f(k)}{2}=f(k+2)$$

$$=\frac{f(k+1)+f(k)}{2}\le\frac{2+2}{2}=2 .$$

Thus, $P(k+2)$ is true. Therefore, by the second principle of mathematical induction, $P(n)$ is true for all $n\in\mathbb{N}$.

1.3A1 Assignment:

(a) Use the principle of mathematical induction to prove that:

$$1^2+3^2+\ldots+(2n-1)^2=\frac{n(2n-1)(2n+1)}{3} \quad \text{for all } n\in\mathbb{N}.$$

(b) Use the modified principle of mathematical induction to prove that:

$n!>2^n$ for all $n\in\mathbb{N}$, $n\ge 4$.

(c) Let $f:\mathbb{N}\to\mathbb{N}$ be defined by $f(1)=5$, $f(2)=13$, and

$$f(n+2)=f(n+1)+2f(n) \quad \text{for all } n\in\mathbb{N}.$$

Prove that: $f(n)=3\cdot 2^n+(-1)^n$ for all $n\in\mathbb{N}$.

(d) Let $f:\mathbb{N}\to\mathbb{N}$ be defined by $f(1)=1$, $f(2)=4$, and

$$f(n+2)=2f(n+1)-f(n)+2 \quad \text{for all } n\in\mathbb{N}.$$

Determine a formula for $f(n)$ and use induction to prove your conclusion.

1.4 The Real Number System

In this section, we will consider the concept of the least upper bound of an ordered field and introduce the real number system with its various properties.

1.4D1 Definition: Field

A **field** is a set F with at least two distinct elements along with two binary operations, addition $(+)$ and multiplication $(\cdot)$, which satisfy the following eleven (11) axioms for addition and multiplication:

(A1) $a,b \in F \Rightarrow a+b \in F.$

(A2) $a,b \in F \Rightarrow a+b = b+a.$

(A3) $a,b,c \in F \Rightarrow a+(b+c) = (a+b)+c.$

(A4) There exists $0 \in F$ such that $a+0=a$ for all $a \in F$; 0 is called the **additive identity** or **zero.**

(A5) $a \in F \Rightarrow$ There exists $-a \in F$ such that $a+(-a)=0$; $-a$ is called the **additive inverse** of a.

(M1) $a,b \in F \Rightarrow a \cdot b \in F.$

(M2) $a,b \in F \Rightarrow a \cdot b = b \cdot a.$

(M3) $a,b,c \in F \Rightarrow a \cdot (b \cdot c) = (a \cdot b) \cdot c.$

(M4) There exists $1 \in F$ such that $1 \neq 0$ and $a \cdot 1 = a$ for all $a \in F$; 1 is called the **multiplicative identity** or **unity** or **one**.

(M5) $a \in F,\ a \neq 0 \Rightarrow$ There exists $a^{-1} \in F$ such that $a \cdot a^{-1} = 1$; a^{-1} is called the **multiplicative inverse** of a.

(D1) $a,b,c \in F \Rightarrow a \cdot (b+c) = a \cdot b + a \cdot c.$

Note: The product $a \cdot b$ is usually written as ab. The sum $a+(-b)$ is usually written as a difference $a-b$. The product $a \cdot b^{-1}$ is usually written as a quotient $\frac{a}{b} = a \cdot \frac{1}{b} = ab^{-1}$.

1.4T1 Theorem:

Let F be a field and $x, y, z \in F$. Then the following hold:

(a) $x+z = y+z \Rightarrow x = y.$

(b) 0 is unique.

(c) Additive inverses are unique.

(d) If $z \neq 0$, then $xz = yz \Rightarrow x = y.$

(e) 1 is unique.

(f) Multiplicative inverses are unique.

(g) $0 \cdot x = 0$

(h) $(-x)(-y) = xy.$

(i) $-x = (-1) \cdot x.$

Proof:

(a) $$x+z=y+z \Rightarrow (x+z)+(-z)=(y+z)+(-z)$$

$$\Rightarrow x+(z+(-z))=y+(z+(-z))$$
$$\Rightarrow x+(0)=y+(0)$$
$$\Rightarrow x=y.$$

(b) Let 0 and $0'$ be two additive identities. Then

$$0+0'=0 \text{ and } 0'+0=0'.$$

But

$$0+0'=0'+0.$$

Hence result.

(c) Let x and y be additive inverses of z. Then

$$z+x=0=z+y, \text{ i.e., } x+z=y+z.$$

But by part (a), $x=y$. Hence result.

(g) $$(0+1)x=1\cdot x \Rightarrow 0\cdot x+1\cdot x=1\cdot x$$

$$\Rightarrow 0\cdot x+x=x$$
$$\Rightarrow 0\cdot x+x=0+x$$

Result follows using part (a).

(h) $$0=0\cdot y=(x+(-x))(-y)=x(-y)+(-x)(-y)$$
$$\Rightarrow (-x)(-y)=-x(-y).$$
$$0=x\cdot 0=x(y+(-y))=xy+x(-y)$$
$$\Rightarrow xy=-x(-y).$$

Hence result.

(i) Let $y=-1$. Then $-y=1$. Use the previous part with this y, to obtain

$$(-1)\cdot x=x\cdot(-1)=(-x)\cdot 1=-x.$$

The proofs of (d), (e), (f) are similar to the proofs of (a), (b), (c) and left as an exercise.

1.4D2 Definition: Ordered Field

A field F is said to be an **ordered field** if it contains a nonempty subset P that satisfies the following order properties:

(O1) $a, b \in P \Rightarrow a + b \in P$ and $ab \in P$.

(O2) If $a \in F$, then one and only one of the following hold:

$$a \in P, \qquad -a \in P, \qquad a = 0.$$

Notation: If $a \in P$, then we write $a > 0$. If $a \notin P$, $a \neq 0$, then we write $a < 0$. $a \geq 0$ means that either $a > 0$ or $a = 0$. $a \leq 0$ means that either $a < 0$ or $a = 0$. If $x, y \in F$, then $x > y$ means that $(x - y) \in P$, i.e., $x - y > 0$. $x \geq y$ means that $x > y$ or $x = y$. $x < y$ means that $(y - x) \in P$, i.e., $x - y < 0$. $x \leq y$ means that $x < y$ or $x = y$.

1.4T2 Theorem:

Let F be an ordered field and $x, y, z \in F$. Then the following hold:

(a) $x \leq y$ or $y \leq x$.

(b) $x \leq y$ and $y \leq x \Rightarrow x = y$.

(c) $x < y$ and $y < z \Rightarrow x < z$. ($<$ may be replaced by $\leq$.)

(d) $x < y \Rightarrow x + z < y + z$. ($<$ may be replaced by $\leq$.)

(e) $x < y$ and $z > 0 \Rightarrow xz < yz$. ($<$ may be replaced by $\leq$.)

(f) $x < y$ and $z < 0 \Rightarrow xz > yz$.

(g) $1 > 0$.

(h) $x \neq 0 \Rightarrow x^2 > 0$.

(i) $x > 0 \Rightarrow x^{-1} > 0$ and $x < 0 \Rightarrow x^{-1} < 0$.

Proof:

(a) Consider the difference $x - y$. The result follows if we use property (O2).

(b) By (O2), $x < y$ and $y \leq x$ is not possible. Therefore, $x = y$.

(c)
$$\begin{aligned}
&(y - x), (z - y) \in P \\
&\Rightarrow ((y - x) + (z - y)) \in P \\
&\Rightarrow (z - x) \in P \Rightarrow x < z.
\end{aligned}$$

(g) $1 < 0 \Rightarrow (-1) > 0 \Rightarrow 1 = 1 \cdot 1 = (-1)(-1) > 0$.
This is a contradiction. Therefore, $1 > 0$. Furthermore, $-1 < 0$.

(h) $x > 0 \Rightarrow x \cdot x > 0.$

$$x < 0 \Rightarrow -x = (-1) \cdot x > 0 \Rightarrow x^2 = x \cdot x = (-x) \cdot (-x) > 0.$$

(i) $x > 0$ and $x^{-1} < 0 \Rightarrow 1 = x(x^{-1}) < 0.$

This is not possible. Therefore, $x^{-1} > 0$. The second part follows easily.

The proofs of (d), (e), (f) follow quite easily and are left as an exercise.

1.4D3 Definition: Absolute Value

Let F be an ordered field and $x \in F$. The absolute value of x, denoted by $|x|$, is defined as follows:

$$|x| = \begin{cases} x & \text{if } x \geq 0 \\ -x & \text{if } x < 0. \end{cases}$$

1.4D4 Definition: Bounded Sets and Upper and Lower Bounds

Let E be a subset of an ordered field F. E is said to be **bounded above** if there exists $\beta \in F$ such that

$$x \leq \beta \text{ for every } x \in E.$$

Such a β is called an **upper bound** of E. E is said to be **bounded below** if there exists $\alpha \in F$ such that

$$x \geq \alpha \text{ for every } x \in E.$$

Such an α is called a **lower bound** of E. E is said to be **bounded** if it is both bounded above and bounded below. Equivalently, E is bounded if there exists $\alpha \in F$, $\alpha > 0$ such that

$$|x| \leq \alpha \text{ for every } x \in E.$$

E is said to be **unbounded** if it is not bounded.

1.4D5 Definition: Least Upper Bound and Greatest Lower Bound

Let E be a nonempty subset of an ordered field F that is bounded above. $\alpha \in F$ is said to be the **least upper bound** or **supremum** of E if the following conditions are satisfied:

(i) α is an upper bound of E.

(ii) If $\beta \in F$ satisfies $\beta < \alpha$, then β is not an upper bound of E.

Note: Condition (ii) is equivalent to $\alpha \leq \beta$ for all upper bounds β of E.

$\alpha \in F$ is said to be the **greatest lower bound** or **infimum** of E if the following conditions are satisfied:

(i) α is a lower bound of E.

(ii) If $\beta \in F$ satisfies $\beta > \alpha$, then β is not a lower bound of E.

Note: Condition (ii) is equivalent to $\alpha \geq \beta$ for all lower bounds β of E.

Notation: The least upper bound of E is denoted by: $\sup E$. The greatest lower bound of E is denoted by: $\inf E$.

1.4T3 Theorem:

Let E be a nonempty subset of an ordered field F that is bounded above. An upper bound α of E is the supremum of E if and only if for every $\beta < \alpha$, there exists $x \in E$ such that $\beta < x \leq \alpha$. Equivalently, for every $\varepsilon > 0$, there exists $x \in E$ such that $\alpha - \varepsilon < x \leq \alpha$ if and only if $\sup E = \alpha$.

Proof: Suppose $\alpha = \sup E$. If $\beta < \alpha$, then β is not an upper bound of E. Therefore, there exists $x \in E$ such that $x > \beta$. Since α is an upper bound of E, $x \leq \alpha$.

Conversely, if α is an upper bound of E satisfying the stated condition, then every $\beta < \alpha$ is not an upper bound of E. Thus $\alpha = \sup E$.

Note: It is easily seen that for every $\varepsilon > 0$, there exists $x \in E$ such that $\alpha \leq x < \alpha + \varepsilon$ implies $\inf E = \alpha$. Furthermore, if a lower bound belongs to E, then it is indeed the infimum of E. If an upper bound belongs to E, then it is the supremum of E.

1.4D6 Definition: The Completeness Axiom

The ordered field F is said to be complete if it satisfies the **completeness axiom**:

(L) Every nonempty subset of F that is bounded above has a supremum in F.

Note: The above axiom is also called the **least upper bound property**. It is equivalent to the **greatest lower bound property**:

Every nonempty subset of F that is bounded below has a greatest lower bound in F.

1.4D7 Definition: The Real Number System

The **real number system** is an ordered field $\mathbb{R}$, that contains the ordered field of rational numbers, and satisfies the completeness axiom (L).

1.4E1 Example: Prove that if $y \in \mathbb{R}$ and $y > 0$, then there exists a unique $\alpha \in \mathbb{R}$ and $\alpha > 0$, such that $\alpha^2 = y$.

Consider the case $y > 1$. Let $C = \{x \in \mathbb{R} \mid x > 0 \text{ and } x^2 < y\}$.

$$y > 1 \Rightarrow y = y \cdot 1 > 1 \cdot 1 = 1^2.$$

Thus $1 \in C$ and C is not empty.

$$y > 1 \Rightarrow y^2 = y \cdot y > y \cdot 1 = y \Rightarrow y \notin C.$$

If $x \in C$ and $y < x$, then

$$y < y^2 < x^2 \Rightarrow y \in C.$$

This is a contradiction. Therefore, $y \geq x$ for every $x \in C$ and y is an upper bound of C. Hence, C has a least upper bound in $\mathbb{R}$. Let $\alpha = \sup C$. Define the real number β by:

(a) $\beta = \alpha + (y - \alpha^2)(\alpha + y)^{-1} = y(\alpha + 1)(\alpha + y)^{-1}$.

Then

(b) $\beta^2 - y = y(y-1)(\alpha^2 - y)(\alpha + y)^{-2}$.

If $\alpha^2 < y$, then by (a) $\beta > \alpha$, and by (b) $\beta^2 < y$. This contradicts that α is an upper bound of C.

If $\alpha^2 > y$, then by (a) $\beta < \alpha$, and by (b) $\beta^2 > y$. If $x \in \mathbb{R}$ and $x \geq \beta$ then $x^2 > y$. Therefore, β is an upper bound of C. This contradicts that $\alpha = \sup C$. It follows that $\alpha^2 = y$. This proves the existence of α. The uniqueness of α is fairly easy to establish and is left as an exercise.

If $y = 1$, then $1^2 = 1$. Thus $\alpha = 1$. The uniqueness is quite obvious. The case $0 < y < 1$ can be handled by using the multiplicative inverse of y.

Note: The unique real number $\alpha \in \mathbb{R}$ and $\alpha > 0$, satisfying $\alpha^2 = y$, is called the square root of y and is denoted by $y^{1/2}$ or $\sqrt{y}$.

1.4D8 Definition: The symbols ∞ and $-\infty$

Let E be a nonempty subset of $\mathbb{R}$. Then we define

(a) $\sup E = \infty$ if E is not bounded above.

(b) $\inf E = -\infty$ if E is not bounded below.

The symbol ∞ is called infinity and the symbol $-\infty$ is called negative infinity. If $x \in \mathbb{R}$, then we adopt the convention:

$$-\infty < x < \infty.$$

Note: For the empty set every element of $x \in \mathbb{R}$, is an upper bound of ϕ. Therefore, we take $\sup \phi = -\infty$. Similarly, $\inf \phi = \infty$. However, it is easily seen that for any nonempty subset E of $\mathbb{R}$,

$$-\infty \leq \inf E \leq \sup E \leq \infty.$$

1.4D9 Definition: Intervals in $\mathbb{R}$

A subset J of $\mathbb{R}$ is called an interval if the following condition is satisfied:

$$x, y \in J \text{ and } x < y \Rightarrow \text{Every } t \text{ satisfying } x < t < y \text{ is in } J.$$

Notation: Let $a, b \in \mathbb{R}$, $a \leq b$. We define all the possible intervals in $\mathbb{R}$.

(a) The **open interval:** $(a,b) = \{x \in \mathbb{R} \mid a < x < b\}$.

(b) The **closed interval:** $[a,b] = \{x \in \mathbb{R} \mid a \leq x \leq b\}$.

(c) The **half-open (half-closed) interval:** $[a,b) = \{x \in \mathbb{R} \mid a \leq x < b\}$.

(d) The **half-open (half-closed) interval:** $(a,b] = \{x \in \mathbb{R} \mid a < x \leq b\}$.

(e) The **infinite open interval:** $(a,\infty) = \{x \in \mathbb{R} \mid a < x < \infty\}$.

(e) The **infinite open interval:** $(-\infty,b) = \{x \in \mathbb{R} \mid -\infty < x < b\}$.

(e) The **infinite closed interval:** $[a,\infty) = \{x \in \mathbb{R} \mid a \leq x < \infty\}$.

(e) The **infinite closed interval:** $(-\infty,b] = \{x \in \mathbb{R} \mid -\infty < x \leq b\}$.

(h) The **real line:** $\mathbb{R} = (-\infty, \infty)$.

1.4T4 Theorem: Archimedian Property

If $x, y \in \mathbb{R}$ and $x > 0$, then there exists a positive integer n such that $nx > y$.

Proof: If $y \leq 0$, then the result is true for all positive integers n. Suppose $y > 0$ and define

$$A = \{nx \mid n \in \mathbb{N}\}.$$

Assume the result is false. Then

$$nx \leq y \text{ for all } n \in \mathbb{N}.$$

Thus y is an upper bound for A, and since A is not empty it has a least upper bound in $\mathbb{R}$. Let

$$\alpha = \sup A.$$

Since $x > 0$,

$$\alpha - x < \alpha.$$

This means that $\alpha - x$ is not an upper bound of A. Therefore, there exists $n \in \mathbb{N}$ such that

$$\alpha - x < nx.$$

Then

$$\alpha < (n+1)x,$$

which contradicts the fact that $\alpha = \sup A$. Therefore, there exists a positive integer n such that

$$nx > y.$$

Note: If we take $y = 1$ and $x = \varepsilon > 0$, then the result reads:

If $\varepsilon > 0$, then there exists a positive integer n such that $\frac{1}{n} < \varepsilon$.

1.4E2 Example:

Let $E=\left\{x=1-\frac{1}{n}\middle| n\in\mathbb{N}\right\}$. Prove that $\inf E=0$ and $\sup E=1$. Do these extreme values belong to *E*? Justify!

$$x\in E \Rightarrow x=1-\frac{1}{n}<1.$$

Thus 1 is an upper bound of *E* and *E* must have a supremum. Let $\varepsilon>0$. Then there exists a positive integer *n* such that $\frac{1}{n}<\varepsilon$. If $x=1-\frac{1}{n}$, then $x\in E$ and

$$1-\varepsilon<x<1.$$

Thus, $\sup E=1$.

$$1-\frac{1}{1}=0\in E \text{ and } 0\le 1-\frac{1}{n} \text{ for all } n\in\mathbb{N}.$$

It follows that $\inf E=0$.

1.4T5 Theorem: Rational Numbers are Dense in $\mathbb{R}$

Suppose $x,y\in\mathbb{R}$ and $x<y$. Then there exists a rational number r such that $x<r<y$.

Proof: Consider the case $x\ge 0$. Then $y-x>0$ and by the previous theorem there exists a positive integer *n* such that

$$n(y-x)>1 \text{ or } ny>1+nx.$$

Furthermore, by the same theorem, we see that the set $\{k\in\mathbb{N}\mid k>nx\}$ is nonempty. Thus, by the well-ordering principle, this set has a smallest element *m*. This implies that

$$m-1\le nx<m.$$

Therefore,

$$nx<m\le 1+nx<ny.$$

Multiplying by n^{-1}, i.e., dividing by *n*, we obtain

$$x<\frac{m}{n}<y.$$

If $x<0<y$, then the result is obvious. If $x<y\le 0$, then $0\le -y<-x$, and there exists positive integers *m* and *n* such that

$$-y<\frac{m}{n}<-x.$$

This implies that

$$x<\frac{-m}{n}<y.$$

1.4T6 Theorem:

For every $x \in \mathbb{R}$, $x > 0$ and for every $n \in \mathbb{N}$, there exists a positive real number y such that $y^n = x$.

Proof: The existence of such a y is deferred to Chapter 3. It is addressed after the Intermediate Value Theorem is presented as it is an immediate consequence of that theorem. The uniqueness of y can be easily established by using mathematical induction to prove that

$$0 < y_1 < y_2 \Rightarrow y_1^n < y_2^n \text{ for all } n \in \mathbb{N}.$$

Note: The unique real number y is written as $x^{1/n}$ or $\sqrt{n}$ and is called the nth root of x.

1.4T6C Corollary:

If a, b are positive real numbers and n is a positive integer, then

$$(ab)^{1/n} = a^{1/n}b^{1/n}.$$

Proof: Let $\alpha = a^{1/n}$ and $\beta = b^{1/n}$. Then

$$ab = \alpha^n \beta^n = (\alpha\beta)^n.$$

By uniqueness, $\alpha\beta = (ab)^{1/n}$.

1.4A1 Assignment:

(a) Let $a, b \in \mathbb{R}$. Prove that $ab = 0 \Rightarrow a = 0 \ or \ b = 0$.

(b) Prove that $|a+b| \le |a| + |b|$ for all $a, b \in \mathbb{R}$.

(c) Let E be a nonempty subset of $\mathbb{R}$ that is bounded below and let L be the set of all lower bounds of E. Prove that $\inf E = \sup L$.

(d) Let E be a nonempty subset of $\mathbb{R}$ and $-E = \{-x \mid x \in E\}$. Prove that $\sup E = -\inf(-E)$.

(e) Let A and B be bounded subsets of $\mathbb{R}$ and define $A + B$ and $A \cdot B$ as follows:

$$A + B = \{a + b \mid a \in A,\ b \in B\} \qquad A \cdot B = \{ab \mid a \in A,\ b \in B\}.$$

(i) Prove that $A \cup B$ is bounded and $\inf(A \cup B) = \min\{\inf A, \inf B\}$.

(ii) Prove that $A + B$ is bounded and $\inf(A + B) = \inf A + \inf B$.

(iii) If A and B contain only positive numbers then prove that $A \cdot B$ is bounded and $\inf(A \cdot B) = \inf A \cdot \inf B$.

(f) Let $f : A \to \mathbb{R}$ and $g : A \to \mathbb{R}$ where $A \subseteq \mathbb{R}$. Prove that

$$\sup\{f(x)+g(x) \mid x\in A\} \le \sup\{f(x) \mid x\in A\} + \sup\{g(x) \mid x\in A\}.$$

1.5 Countable and Uncountable Sets

In this section, we take a closer look at infinite sets and the notion of countability and uncountability.

1.5D1 Definition: Equivalent Sets

Two sets A and B are said to be **equivalent** (or have the same **cardinality**) if there exists $f: A \to B$ such that f is one-to-one and onto. In this case, we write $A \sim B$. It is easily seen that the notion of equivalence of sets satisfies the following:

(i) $A \sim A$. (reflexive)

(ii) $A \sim B \Leftrightarrow B \sim A$. (symmetric)

(iii) $A \sim B$ and $B \sim C \Rightarrow A \sim C$. (transitive)

1.5D2 Definition: Finite, Countable, and Uncountable Sets

Let $n \in \mathbb{N}$ and define $\mathbb{N}_n = \{1, 2, \ldots, n\}$. Suppose A is a set.

(a) A is **finite** if $A \sim \mathbb{N}_n$ for some $n \in \mathbb{N}$, or if $A = \phi$.
(b) A is **infinite** if is not finite.
(c) A is **countable** if $A \sim \mathbb{N}$.
(d) A is **uncountable** if A is infinite and not countable.
(e) A is **at most countable** if A is finite or countable.

Note: Countable sets are sometimes called **denumerable** sets or **enumerable** sets.

1.5T1 Theorem

If A is countable and $A \sim B$, then B is countable.

Proof: $A \sim B \Rightarrow B \sim A$; A is countable $\Rightarrow A \sim \mathbb{N}$. Therefore, $B \sim \mathbb{N}$. Hence, result.

1.5E1 Example: We will show that $\mathbb{Z}$ is countable by defining $f: \mathbb{Z} \to \mathbb{N}$ as follows:

$$f(m) = \begin{cases} 2m & \text{if } m > 0 \\ -2m+1 & \text{if } m \le 0. \end{cases}$$

Then

$$m > 0 \Rightarrow f(m) \in \mathbb{N}.$$

$$m \le 0 \Rightarrow f(m) = -2m+1 \ge 1 \text{ and } f(m) \in \mathbb{N}.$$

The range of f is contained in the set of natural numbers. We will first show that f is onto. If $n \in \mathbb{N}$ and n is odd, let $m = -(n-1)/2$. Then

$$m \le 0 \text{ and } f(m) = -2m+1 = (n-1)+1 = n.$$

If n is even, let $m = n/2$. Then

$$m > 0 \text{ and } f(m) = 2m = n.$$

This proves that f is onto. Now suppose $f(m_1) = f(m_2)$. If both m_1 and m_2 are positive, then

$$2m_1 = 2m_2 \text{ and } m_1 = m_2.$$

If both m_1 and m_2 are nonpositive, then

$$-2m_1 + 1 = -2m_2 + 1 \text{ and } m_1 = m_2.$$

If $m_1 > 0$ and $m_2 \le 0$, then

$$2m_1 = -2m_2 + 1.$$

This implies that

$$2(m_1 + m_2) = 1.$$

This is not possible. Similarly, the case where $m_1 \le 0$ and $m_2 > 0$ is also not possible. Therefore f is one-to-one. This completes the proof that $\mathbb{Z}$ is countable.

1.5T2 Theorem:

$\mathbb{N} \times \mathbb{N}$ is countable.

Proof: Define $g : \mathbb{N} \times \mathbb{N} \to \mathbb{N}$ by

$$g(m,n) = 2^{m-1}(2n-1).$$

Suppose

$$g(m,n) = 2^{m-1}(2n-1) = g(p,q) = 2^{p-1}(2q-1).$$

Then

$$2^{m-p}(2n-1) = (2q-1).$$

Since the right side of the equation is odd,

$$2^{m-q} = 1 \text{ and } m = q.$$

$$2^{m-1} = 2^{p-1} \text{ and } n-1 = q-1$$

$$\Rightarrow m = p \text{ and } n = q.$$

Hence, g is one-to-one. Suppose $p \in \mathbb{N}$. We can write

$$p = 2^{m-1}(2n-1) \text{ for some } m, n \in \mathbb{N}.$$

(Factor out 2^{m-1} so that the remaining factor is odd; this can be written as $2n-1$.) Then

$$g(m,n) = 2^{m-1}(2n-1) = p.$$

Thus, g is onto. This shows that $\mathbb{N}\times\mathbb{N}\sim\mathbb{N}$ and completes the proof.

1.5D3 Definition: Sequences

Let A be an arbitrary set of objects. Then a function $f:\mathbb{N}\to A$ is called a **sequence** in A. For each $n\in\mathbb{N}$, let $x_n = f(n)$. Then x_n is called the nth **term** of the sequence f.

Notation: Sequences are usually denoted by $\{x_n\}_{n=1}^{\infty}$ or simply $\{x_n\}$, rather than the function f. Note that the sequence $\{x_n\}_{n=1}^{\infty}$ is different from the set of values of the sequence $\{x_n \mid n\in\mathbb{N}\}$. For example, $\left\{(-1)^n\right\}_{n=1}^{\infty}$ denotes the sequence f where $f(n)=(-1)^n$, while $\{x_n \mid n\in\mathbb{N}\}=\{-1,1\}$.

Note: If A is countable, then there exists a one-to-one function f from $\mathbb{N}$ onto A. Therefore, $A = Ran(f)=\{x_n \mid n\in\mathbb{N}\}$, and $\{x_n \mid n\in\mathbb{N}\}$ is called an **enumeration** of the set A.

1.5T3 Theorem

Every infinite subset of a countable set is countable.

Proof: Let A be a countable set and let $\{x_n \mid n\in\mathbb{N}\}$ be an enumeration of A. Suppose B is an infinite subset of A. Then each element of B is of the form x_k for some $k\in\mathbb{N}$. We construct a function $f:\mathbb{N}\to B$ as follows:

Let n_1 be the smallest positive integer such that $x_{n_1}\in B$. (Well-ordering principle.) Define $f(1)=x_{n_1}$. Having chosen

$$n_1 < n_2 < \ldots < n_k \text{ with } f(j)=x_{n_j};\ j=1,2,\ldots,k,$$

let n_{k+1} be the smallest integer such that

$$x_{n_{k\ 1}} \in B\setminus\left\{x_{n_1},x_{n_2},\ldots,x_{n_k}\right\}.$$

Let

$$f(k+1)=x_{n_{k+1}}.$$

Since B is infinite, f is defined on the set of natural numbers. If $k<m$, then

$$n_k < n_m \text{ and } f(k)=x_{n_k}\neq x_{n_m}=f(m).$$

Therefore, f is one-to-one. If $x\in B$, then

$$x = x_j \text{ for some } j.$$

But $j=n_k$ for some k, and

$$f(k)=x_{n_k}=x_j=x.$$

Hence, f is onto.

1.5T4 Theorem

Let $f:\mathbb{N}\to A$ be onto. Then A is at most countable.

Proof: If A is finite, then it is at most countable. Suppose A is infinite. Since f is onto, if $a\in A$, then there exists $n\in\mathbb{N}$ such that $f(n)=a$. Using the well-ordering principle, we see that for each $a\in A$,

$$f^{-1}(\{a\})=\{n\in\mathbb{N}\mid f(n)=a\}$$

has a smallest element, say n_a. Define a function $g:A\to\mathbb{N}$ by

$$g(a)=n_a.$$

Since f is a function, the two sets

$$f^{-1}(\{a\})=\{n\in\mathbb{N}\mid f(n)=a\}$$

and

$$f^{-1}(\{b\})=\{n\in\mathbb{N}\mid f(n)=b\},$$

do not have any elements in common if $a\neq b$. It is easily seen that

$$g(a)=n_a\neq n_b=g(b)$$

if $a\neq b$. Thus, g is one-to-one and

$$g(A)=\{g(a)\mid a\in A\}$$

is an infinite subset of $\mathbb{N}$. Therefore, $A\sim g(A)$ and by 1.5T3, $g(A)$ is countable. Now we use 1.5T1 to conclude that A is countable.

1.5D4 Definition: Indexed Families of Sets

Let A and X be nonempty sets. An **indexed family** of subsets of X with index set A is a function from A into the power set $\mathrm{P}(X)$ of X.

Notation: If $f:A\to\mathrm{P}(A)$, then for each $\alpha\in A$, let $E_\alpha=f(\alpha)$. As in the case for sequences, we denote this function by $\{E_\alpha\}_{\alpha\in A}$. If $A=\mathbb{N}$, then $\{E_n\}_{n\in\mathbb{N}}$ is called a **sequence of subsets** of X. In this instance, we may adopt the notation $\{E_n\}_{n=1}^{\infty}$ or simply $\{E_n\}$.

1.5D5 Definition: Union and Intersection

Let $\{E_\alpha\}_{\alpha\in A}$ be an **indexed family** of subsets of X.

The **union** of this family is defined by

$$\bigcup_{\alpha\in A}E_\alpha=\{x\in X\mid x\in E_\alpha \text{ for some } \alpha\in A\}.$$

The **intersection** of this family is defined by

$$\bigcap_{\alpha \in A} E_\alpha = \{x \in X \mid x \in E_\alpha \text{ for every } \alpha \in A\}.$$

Note: If $A = \mathbb{N}$, we use the notations $\bigcup_{n=1}^{\infty} E_n$ and $\bigcap_{n=1}^{\infty} E_n$ instead of $\bigcup_{\alpha \in A} E_\alpha$ and $\bigcap_{\alpha \in A} E_\alpha$, respectively. Also if $A = \mathbb{N}_k = \{1, 2, ..., k\}$, then $\bigcup_{n=1}^{k} E_n$ and $\bigcap_{n=1}^{k} E_n$ are used instead of $\bigcup_{\alpha \in A} E_\alpha$ and $\bigcap_{\alpha \in A} E_\alpha$.

1.5E2 Example: Find the union and intersection of the family of sets given by:

$$E_n = \{x \in \mathbb{R} | 1 - 1/n \le x < 2 - 1/n\}, \quad n \in \mathbb{N}.$$

In this example, we will use $\bigcup$ and $\bigcap$ in place of $\bigcup_{n=1}^{\infty} E_n$ and $\bigcap_{n=1}^{\infty} E_n$ respectively. The union is the interval $[0,2)$ and the intersection is ϕ.

If $x < 0$, then $x \notin E_n$ for any $n \in \mathbb{N}$ and $x \notin \bigcup$.

$$E_1 \cup E_2 = [0,1) \cup [1/2, 3/2) = [0, 3/2).$$

Therefore, if $0 \le x \le 1$, then $x \in \bigcup$.

Suppose $1 < x < 2$. Then there exists $n \in \mathbb{N}$ such that

$$0 < 1/n < 2 - x \Rightarrow 1 < x < 2 - 1/n \text{ and } x \in E_n.$$

Therefore, $x \in \bigcup$.

If $x \ge 2$, then $x \notin E_n$ for any $n \in \mathbb{N}$. In this case, $x \notin \bigcup$. This shows that

$$\bigcup_{n=1}^{\infty} E_n = [0,2).$$

Since $\bigcap \subseteq \bigcup$, we have to consider the interval $[0,2)$ only. Suppose $0 \le x < 1$. Then there exists $n \in \mathbb{N}$ such that

$$0 < 1/n < 1 - x \Rightarrow 0 \le x < 1 - 1/n \text{ and } x \notin E_n \Rightarrow x \notin \bigcap.$$

If $x \ge 1$, then $x \notin E_1 = [0,1) \Rightarrow x \notin \bigcap$. This shows that $\bigcap_{n=1}^{\infty} E_n = \phi$.

1.5A1 Assignment: Find the intersection and the union of the family of sets given by

(a) $E_n = \{x \in \mathbb{R} | 1 + 1/n < x < 2 - 1/n\}, \quad n \in \mathbb{N}.$

(b) $E_n = \{x \in \mathbb{R} | 1 - 1/n < x < 2 + 1/n\}, \quad n \in \mathbb{N}.$

(c) $E_n = \{x \in \mathbb{R} | 1 - 1/n \le x \le 2 + 1/n\}, \quad n \in \mathbb{N}.$

Note: The following results are easily established and the proofs are left as an exercise.

1.5T5 Theorem: Distributive Laws

Let E and $\{E_\alpha\}_{\alpha \in A}$ be subsets of a set X. Then

(a) $E \cap \bigcup_{\alpha \in A} E_\alpha = \bigcup_{\alpha \in A} (E \cap E_\alpha)$.

(b) $E \cup \bigcap_{\alpha \in A} E_\alpha = \bigcap_{\alpha \in A} (E \cup E_\alpha)$.

1.5T6 Theorem: DeMorgan's Laws

Let E and $\{E_\alpha\}_{\alpha \in A}$ be subsets of a set X. Then

(a) $E \setminus \bigcup_{\alpha \in A} E_\alpha = \bigcap_{\alpha \in A} (E \setminus E_\alpha)$.

(b) $E \setminus \bigcap_{\alpha \in A} E_\alpha = \bigcup_{\alpha \in A} (E \setminus E_\alpha)$.

Note: $E \setminus \bigcap_{\alpha \in A} E_\alpha$, $E \setminus \bigcup_{\alpha \in A} E_\alpha$, and $(E \setminus E_\alpha)$ can be replaced by $\left(\bigcap_{\alpha \in A} E_\alpha\right)^c$, $\left(\bigcup_{\alpha \in A} E_\alpha\right)^c$, and $E_\alpha{}^c$.

1.5T7 Theorem:

Let $\{X_\alpha\}_{\alpha \in A}$ be a family of subsets of a set X, $\{Y_\alpha\}_{\alpha \in A}$ be a family of subsets of a set Y, and let $f : X \to Y$. Then

(a) $f\left(\bigcup_{\alpha \in A} X_\alpha\right) = \bigcup_{\alpha \in A} f(X_\alpha)$.

(b) $f\left(\bigcap_{\alpha \in A} X_\alpha\right) \subseteq \bigcap_{\alpha \in A} f(X_\alpha)$.

(c) $f^{-1}\left(\bigcup_{\alpha \in A} Y_\alpha\right) = \bigcup_{\alpha \in A} f^{-1}(Y_\alpha)$.

(d) $f^{-1}\left(\bigcap_{\alpha \in A} Y_\alpha\right) = \bigcap_{\alpha \in A} f^{-1}(Y_\alpha)$.

Proof: (a) $y \in f\left(\bigcup_{\alpha \in A} X_\alpha\right) \Leftrightarrow$ there exists $x \in \bigcup_{\alpha \in A} X_\alpha$ such that $y = f(x)$

$\Leftrightarrow y = f(x)$ for $x \in X_\alpha$ for some $\alpha \in A$

$\Leftrightarrow y \in f(X_\alpha)$ for some $\alpha \in A \Leftrightarrow y \in \bigcup_{\alpha \in A} f(X_\alpha)$.

(b)
$$y \in f\left(\bigcap_{\alpha\in A} X_\alpha\right) \Rightarrow y = f(x) \text{ for some } x \in \bigcap_{\alpha\in A}(X_\alpha)$$
$$\Rightarrow y = f(x) \text{ for some } x \in X_\alpha \text{ for all } \alpha \in A$$
$$\Rightarrow y \in f(X_\alpha) \text{ for all } \alpha \in A \Rightarrow y \in \bigcap_{\alpha\in A} f(X_\alpha).$$

(c)
$$Y_\alpha \subseteq \bigcup_{\alpha\in A} Y_\alpha \text{ for all } \alpha \in A \Rightarrow f^{-1}(Y_\alpha) \subseteq f^{-1}\left(\bigcup_{\alpha\in A} Y_\alpha\right) \text{ for all } \alpha \in A$$
$$\Rightarrow \bigcup_{\alpha\in A} f^{-1}(Y_\alpha) \subseteq f^{-1}\left(\bigcup_{\alpha\in A} Y_\alpha\right).$$
$$x \in f^{-1}\left(\bigcup_{\alpha\in A} Y_\alpha\right) \Rightarrow f(x) \in \bigcup_{\alpha\in A} Y_\alpha$$
$$\Rightarrow f(x) \in Y_\alpha \text{ for some } \alpha \in A$$
$$\Rightarrow x \in f^{-1}(Y_\alpha) \text{ for some } \alpha \in A \Rightarrow x \in \bigcup_{\alpha\in A} f^{-1}(Y_\alpha).$$

Thus $f^{-1}\left(\bigcup_{\alpha\in A} Y_\alpha\right) = \bigcup_{\alpha\in A} f^{-1}(Y_\alpha)$.

(d) Exercise.

1.5T8 Theorem:

Let $\{E_n\}_{n=1}^{\infty}$ be a sequence of countable sets and $E = \bigcup_{n=1}^{\infty} E_n$. Then *E* is countable.

Proof: Since E_n is countable for each $n \in \mathbb{N}$, we can write

$$E_n = \{x_{n,k} \mid k = 1,\ 2, \ldots\}.$$

Now define a function $f : \mathbb{N}\times\mathbb{N} \to E$ by

$$f(n,k) = x_{n,k}.$$

This is a function from $\mathbb{N}\times\mathbb{N}$ onto *E*. Since $\mathbb{N}\times\mathbb{N}$ is countable, there exists $g : \mathbb{N} \to \mathbb{N}\times\mathbb{N}$ such that *g* is one-to-one and onto. It follows that $g \circ f$ is onto (1.5T4) and since *E* is clearly not finite, *E* must be countable.

1.5T8C Corollary: $\mathbb{Q}$ is countable.

Proof: Define $E_n = \left\{\frac{m}{n}\,\middle|\, m \in \mathbb{Z}\right\}$ for each $n \in \mathbb{N}$. Then E_n is countable and $\bigcup_{n=1}^{\infty} E_n$ is countable by the above theorem. But $\bigcup_{n=1}^{\infty} E_n = \mathbb{Q}$. Hence result.

Note: We have so far shown that $\mathbb{N}$ and $\mathbb{Q}$ are countable. We know that there are real numbers that are not rational, e.g., $\sqrt{2}$. We would like to know whether $\mathbb{R}$ is countable or not. $\mathbb{R}$ is clearly infinite because it contains both $\mathbb{N}$ and $\mathbb{Q}$.

1.5D5 Definition: BinaryExpansion of a Real Number in $[0,1]$

The **binary expansion** of 0 is defined to be 0. Suppose $x \in \mathbb{R},\ 0 < x \le 1$. Let $n_1 \in \{0,1\}$ be the largest integer such that $n_1 / 2 \le x$. Having chosen $n_1, n_2, ..., n_k$ let $n_{k+1} \in \{0,1\}$ be the largest integer such that

$$\frac{n_1}{2} + \frac{n_2}{2} + ... + \frac{n_{k+1}}{2} < x.$$

The expression $.n_1 n_2 n_3 ...$ is called the **binary expansion** of x.

Note: It can be shown that if

$$E = \left\{ \frac{n_1}{2} + \frac{n_2}{2} + ... + \frac{n_k}{2} \,\middle|\, k \in \mathbb{N} \right\},$$

then

$$\sup E \quad x.$$

1.5T8 Theorem:

The closed interval $[0,1]$ is uncountable.

Proof: Since $[0,1] \cap \mathbb{Q}$ is infinite, $[0,1]$ must be infinite. Suppose $[0,1]$ is countable. Then $[0,1] = \{x_n \mid n \in \mathbb{N}\}$, where each $x_n \in [0,1]$ has a binary expansion $x_n = .x_{n_1} x_{n_2} x_{n_3} ...$ Define $y = .y_1 y_2 y_3 ...,$ where

$$y_n = \begin{cases} 1 & \text{if } x_{n_n} = 0 \\ 0 & \text{if } x_{n_n} = 1. \end{cases}$$

Then $y \in [0,1]$ and $y \ne x_n$ for any $n \in \mathbb{N}$. This is a contradiction. Therefore, $[0,1]$ is uncountable.

1.5T8C Corollary:

$\mathbb{R}$ is uncountable.

Proof: Since $[0,1]$ is uncountable and $[0,1] \subseteq \mathbb{R}$, $\mathbb{R}$ cannot be countable.

1.5A2 Assignment:

(a) Prove that the set of odd positive integers is countable.

(b) Prove that $(0,1) \sim (0,\infty)$.

(c) Prove that $[0,1]\times[0,1] \sim [0,1]$.

CHAPTER

Sequences and Series of Real Numbers

2.1 Convergent Sequences

We have already introduced the notion of absolute value (of a real number) (1.4D3). In this section, we provide several useful properties of absolute value. Sequences were introduced in (1.5D3); the convergence of sequences and some properties of convergent sequences will also be discussed here.

2.1T1 Theorem:

Let $x, y \in \mathbb{R}$. Then the following hold:

(a) $|-x| = |x|$.

(b) $|xy| = |x||y|$.

(c) $|x| = \sqrt{x^2}$.

(d) If $r > 0$, then $|x| < r \Leftrightarrow -r < x < r$.

(e) $-|x| \le x \le |x|$.

Proof: (a), (b) follow easily by considering all possible cases for x and y, i.e. < 0, $= 0$, > 0.

(c) Suppose $x > 0$. Then by (1.4E1, p12),

$$|x| = x = \left(x^2\right)^{1/2} = \sqrt{x^2}.$$

If $x = 0$, then the result is trivially true. If $x < 0$, then

$$|x| = -x = \left((-x)^2\right)^{1/2} = \left(x^2\right)^{1/2} = \sqrt{x^2}. \qquad \text{(1.4E1, 1.4T1(h))}.$$

(d) and (e) are left as an exercise.

2.1T2 Theorem: The Triangle Inequality

For all $x, y \in \mathbb{R}$, $|x+y| \le |x|+|y|$.

Proof: Let $x, y \in \mathbb{R}$. Then

$$0 \le (x+y)^2 = x^2 + 2xy + y^2 \qquad (1.4\text{T2(h)})$$

$$\le |x|^2 + 2|x||y| + |y|^2 \qquad (2.1\text{T1(c), (e)})$$

$$= (|x|+|y|)^2.$$

By 2.1T1(c),

$$|x+y| = \sqrt{(x+y)^2} \le \sqrt{(|x|+|y|)^2} = |x|+|y|.$$

Note: It is easily seen that if $x, y \ge 0$, then $x \le y \Leftrightarrow x^2 \le y^2$.

2.1T2C Corollary:

Let $x, y, z \in \mathbb{R}$. Then

(a) $\big||x|-|y|\big| \le |x-y|$.

(b) $\big||x|-|y|\big| \le |x+y|$.

(c) $|x-y| \le |x-z|+|z-y|$.

Proof:

(a) $$|x| = |(x-y)+y| \le |x-y|+|y| \Rightarrow |x|-|y| \le |x-y|. \qquad (2.1\text{T2})$$

Therefore,

$$-(|x|-|y|) = |y|-|x| \le |y-x| = |-(x-y)| = |x-y|. \text{ (Switch } x \text{ and } y.)$$

But

$$\big||x|-|y|\big| = \max\{|x|-|y|, -(|x|-|y|)\}.$$

Hence result.

(b) Using (a),

$$\big||x|-|y|\big| = \big||x|-|-y|\big| \le |x-(-y)| = |x+y|.$$

(c) $$|x-y| = |(x-z)+(z-y)| \le |x-z|+|z-y|. \qquad (2.1\text{T2 p 22})$$

2.1E1 Example:

Determine the set of all real numbers x that satisfy the inequality:

(a) $|2x+1| \le 5.$ (b) $|3x-2| > 7.$

Proof:

(a) By (2.1T2(d)),

$$-5 \le 2x+1 \le 5 \Rightarrow -3 \le x \le 2$$

Solution: $[-3,2]$

(b)
$$|3x-2| > 7 \Rightarrow 3x-2 > 7 \text{ or } -(3x-2) > 7$$
$$\Rightarrow x > 3 \text{ or } -3x > -5, \text{ i.e., } x < 5/3$$

Solution: $(-\infty, 5/3) \cup (3, \infty).$

2.1D1 Definition: Neighborhood of a Point

Let $p \in \mathbb{R}$ and $\varepsilon > 0.$ An $\varepsilon-$ **neighborhood** of p is the set, denoted by $N_\varepsilon(p),$ given by:

$$N_\varepsilon(p) = \{x \in \mathbb{R} \mid |x-p| < \varepsilon\} = (p-\varepsilon, p+\varepsilon).$$

A **deleted neighborhood** of p, denoted by $N_\varepsilon'(p),$ is the set given by:

$$N_\varepsilon'(p) = \{x \in \mathbb{R} \mid |x-p| < \varepsilon,\ x \ne p\} = (p-\varepsilon, p+\varepsilon) \setminus \{p\}.$$

2.1E2 Example:

(a) $N_2(-3) = \{x \in \mathbb{R} \mid |x-(-3)| < 2\} = \{x \in \mathbb{R} \mid -5 < x < -1\} = (-5,-1).$

(b) $N_5'(2) = \{x \in \mathbb{R} \mid |x-2| < 5,\ x \ne 2\} = \{x \in \mathbb{R} \mid -3 < x < 7,\ x \ne 2\} = (-3,7) \setminus \{2\}.$

2.1D2 Definition: Convergence of Sequences

Let $\{p_n\}_{n=1}^{\infty}$ be a sequence in $\mathbb{R}.$ $\{p_n\}_{n=1}^{\infty}$ is said to **converge** if there exists $p \in \mathbb{R}$ such that for every $\varepsilon > 0,$ there exists $n_0 \in \mathbb{N}$ such that

$$p_n \in N(p,\varepsilon) \text{ for every } n \ge n_0.$$

In this case, we say that the sequence $\{p_n\}_{n=1}^{\infty}$ **converges** to p, or p is the **limit of the sequence** $\{p_n\}_{n=1}^{\infty}$, and we write:

$$\lim_{n\to\infty} p_n = p \text{ or } p_n \to p \text{ as } n \to \infty \text{ or simply } p_n \to p.$$

If $\{p_n\}_{n=1}^{\infty}$ does not converge, then it is said to **diverge**.

Note: The statement: $p_n \in N(p,\varepsilon)$ for every $n \geq n_0$ is equivalent to the statement:

$$|p_n - p| < \varepsilon \text{ for every } n \geq n_0.$$

2.1E3 Example:

(a) Consider the sequence $\left\{p_n = \frac{1}{n}\right\}_{n=1}^{\infty}$. We will show that this sequence converges to 0.

Let $\varepsilon > 0$ be given and let n_0 be the smallest positive integer such that $n_0 > \frac{1}{\varepsilon}\cdot$ Such an integer exists by the well-ordering principle. Then for all $n \geq n_0$,

$$|p_n - 0| = \left|\frac{1}{n}\right| \leq \frac{1}{n_0} < \varepsilon.$$

Therefore, $\lim\limits_{n\to\infty} p_n = 0.$

(b) The constant sequence $\{p_n = p\}_{n=1}^{\infty}$, where $p \in \mathbb{R}$ converges to p because for any $\varepsilon > 0,$

$$|p_n - p| = |0| < \varepsilon \text{ for all } n \geq 1.$$

(c) Consider the sequence $\left\{p_n = \frac{n-2}{3n-1}\right\}_{n=1}^{\infty}$. We will show that $\lim\limits_{n\to\infty} p_n = \frac{1}{3}$. Let $\varepsilon > 0$ be given and let n_0 be the smallest positive integer such that $n_0 > \frac{1}{\varepsilon}\cdot$ Then

$$\left|\frac{n-2}{3n-1} - \frac{1}{3}\right| = \left|\frac{5}{3(3n-1)}\right|$$

$$\leq \frac{5}{3(3n-n)} = \frac{5}{6n} < \frac{1}{n} < \varepsilon \text{ for all } n \geq n_0.$$

Therefore, $\lim\limits_{n\to\infty} p_n = \frac{1}{3}.$

(d) The sequence $\left\{p_n = \frac{(-1)^n n}{n+1}\right\}_{n=1}^{\infty}$ diverges. Note that since $\frac{n}{n+1} \geq \frac{1}{2}$ for all $n \in \mathbb{N}$,

$$|p_n - p_{n+1}| = \left|\frac{n}{n+1} + \frac{n+1}{n+2}\right|$$

$$\geq \frac{1}{2} + \frac{1}{2} = 1 \text{ for all } n \in \mathbb{N}.$$

Suppose $\lim_{n\to\infty} p_n = p$ for some $p \in \mathbb{R}$. Then there exists $n_0 \in \mathbb{N}$ such that

$$|p_n - p| < \frac{1}{2} \text{ for all } n \geq n_0.$$

But if $n \geq n_0$, then

$$|p_n - p_{n+1}| = |p_n - p| + |p - p_{n+1}| < \frac{1}{2} + \frac{1}{2} = 1.$$

This is a contradiction. Therefore, the sequence must be divergent.

2.1D3 Definition: Bounded Sequences

A sequence $\{p_n\}_{n=1}^{\infty}$ in $\mathbb{R}$ is said to be **bounded** if there exists $M > 0$ such that

$$|p_n| \leq M \text{ for all } n \in \mathbb{N}.$$

This is equivalent to saying that the range of the sequence is a bounded subset of $\mathbb{R}$. An **unbounded** sequence is one that is not bounded.

A sequence $\{p_n\}_{n=1}^{\infty}$ in $\mathbb{R}$ is said to be **bounded above** if there exists $M \in \mathbb{R}$ such that

$$p_n \leq M \text{ for all } n \in \mathbb{N}.$$

A sequence $\{p_n\}_{n=1}^{\infty}$ in $\mathbb{R}$ is said to be **bounded below** if there exists $m \in \mathbb{R}$ such that

$$p_n \geq m \text{ for all } n \in \mathbb{N}.$$

It follows that a sequence $\{p_n\}_{n=1}^{\infty}$ is **bounded** if and only if it is both bounded above and bounded below.

2.1L1 Lemma:

If $|p| < \varepsilon$ for every $\varepsilon > 0$, then $p = 0$.

Proof: Suppose $|p| < \varepsilon$ for every $\varepsilon > 0$ and $p \neq 0$. Then $|p| > 0$. Let $\varepsilon = |p|/2$. This implies that

$$|p| < |p|/2 \Rightarrow 2 < 1.$$

This gives us a contradiction because

$$2 = 1 + 1 \text{ and } 1 > 0.$$

Hence result.

2.1T3 Theorem:

Let $\{p_n\}_{n=1}^{\infty}$ be a convergent sequence in $\mathbb{R}$. Then

(a) $\lim_{n\to\infty} p_n$ is unique.

(b) $\{p_n\}_{n=1}^{\infty}$ is bounded.

Proof:

(a) Suppose $\lim_{n\to\infty} p_n = p$ and $\lim_{n\to\infty} p_n = q$. Let $\varepsilon > 0$ be arbitrary. Then there exists $n_1, n_2 \in \mathbb{N}$ such that

$$|p_n - p| < \varepsilon/2 \text{ for all } n \geq n_1 \text{ and } |p_n - q| < \varepsilon/2 \text{ for all } n \geq n_2.$$

Let $n_0 = \max\{n_1, n_2\}$. Then for all $n \geq n_0$,

$$\begin{aligned} |p-q| &= |p - p_n + p_n - q| \leq |p - p_n| + |p_n - q| \\ &< \varepsilon/2 + \varepsilon/2 = \varepsilon. \end{aligned}$$

But ε was arbitrary. Therefore by (2.1L1),

$$p - q = 0, \text{ i.e., } p = q.$$

(b) Suppose $\lim_{n\to\infty} p_n = p$. Then there exists $n_0 \in \mathbb{N}$ such that

$$|p_n - p| < 1 \text{ for all } n \geq n_0.$$

This implies that

$$|p_n| < 1 + |p| \text{ for all } n \geq n_0$$

because $\big||p_n| - |p|\big| \leq |p_n - p|$ (2.1T2C(a)and 2.1T1(d)). Let

$$M = \max\left\{|p_1|, |p_2|, \ldots, |p_{n_0}|, 1 + |p|\right\}.$$

Then

$$|p_n| \leq M \text{ for all } n \in \mathbb{N}.$$

Therefore, the sequence is bounded.

2.1E4 Example:

(a) The sequence $\{p_n = n\}_{n=1}^{\infty}$ is not bounded in $\mathbb{R}$. Therefore, it cannot be convergent.

(b) The sequence $\left\{p_n = \dfrac{(-1)^n n}{n+1}\right\}_{n=1}^{\infty}$ diverges. However, the sequence is bounded by 1. This shows that the converse of the previous theorem is false. Every bounded sequence need not be convergent.

2.1T4 Theorem:

Suppose $\{p_n\}_{n=1}^{\infty}$ is a real sequence such that $\lim_{n\to\infty} p_n > 0$. Then there exists $n_0 \in \mathbb{N}$ such that $p_n > 0$ for all $n \geq n_0$.

Proof: Suppose $\lim_{n\to\infty} p_n = p > 0$. Then there exists $n_0 \in \mathbb{N}$ such that

$$|p_n - p| < p/2 \text{ for all } n \geq n_0.$$

This implies that

$$0 < p/2 < p_n < 3p/2 \text{ for all } n \geq n_0.$$

Note: A similar result holds when $\lim_{n\to\infty} p_n = p < 0$.

2.1E5 Example:

Prove that if $\lim_{n\to\infty} p_n = p$, then $\lim_{n\to\infty} p_n^3 = p^3$.

Suppose $p = 0$. Let $\varepsilon > 0$ be arbitrary. Then there exists $n_0 \in \mathbb{N}$ such that for all $n \geq n_0$,

$$|p_n - p| = |p_n| < \varepsilon^{1/3}.$$

Therefore, if $n \geq n_0$, then

$$\begin{aligned} |p_n^3 - p^3| &= |p_n^3| = |p_n|^3 \\ &< \left(\varepsilon^{1/3}\right)^3 = \varepsilon. \end{aligned}$$

Now let $p \neq 0$. Then there exists $n_1 \in \mathbb{N}$ such that for all $n \geq n_1$,

$$|p_n - p| < |p|/2.$$

Therefore, for all $n \geq n_1$,

$$\begin{aligned} &\big||p_n| - |p|\big| \leq |p_n - p| < |p|/2 \\ &\Rightarrow |p|/2 < |p_n| < 3|p|/2. \end{aligned}$$

Then for all $n \geq n_1$,

$$\begin{aligned} |p_n^3 - p^3| &= |p_n - p||p_n^2 + p_n p + p^2| \\ &\leq |p_n - p|\left(|p_n^2| + |p_n||p| + |p^2|\right) \\ &\leq |p_n - p|\left(9|p|^2/4 + 3|p|^2/2 + |p^2|\right) \\ &= 19|p|^2|p_n - p|/4. \end{aligned}$$

Let $\varepsilon > 0$ be given. Then there exists $n_2 \in \mathbb{N}$ such that

$$|p_n - p| < \frac{4\varepsilon}{19|p|^2} \text{ for all } n \geq n_2.$$

Let $n_3 = \max\{n_1, n_2\}$. Then for all $n \geq n_3$,

$$|p_n^3 - p^3| \leq 19|p|^2 |p_n - p| / 4 < \varepsilon.$$

2.1E6 Example: Let $\{p_n\}$ be a real sequence satisfying for some $c > 0$,

$$|p_n - p_{n+1}| \geq c \text{ for all } n \in \mathbb{N}.$$

Prove that $\{p_n\}$ diverges.

Suppose $\{p_n\}$ converges to *p*. Then there exists $n_0 \in \mathbb{N}$ such that

$$|p_n - p| < c/2 \text{ for all } n \geq n_0.$$

Then if $n \geq n_0$,

$$\begin{aligned} |p_n - p_{n+1}| &\leq |p_n - p| + |p - p_{n+1}| \\ &< c/2 + c/2 = c. \end{aligned}$$

This is a contradiction. Therefore, $\{p_n\}$ diverges.

2.1A1 Assignment

(a) Determine all $x \in \mathbb{R}$ that satisfy each of the inequalities:

(i) $|x-1| + |x+1| < 4$ (ii) $|x| + |x-1| > 3$.

(b) Prove that $\left\{\dfrac{n^2+2}{2n^2-1}\right\}_{n=1}^{\infty}$ converges to $\dfrac{1}{2}$.

(c) Prove that $\left\{\dfrac{n}{n^2+1}\right\}_{n=1}^{\infty}$ converges to 0.

(d) Prove that $\left\{\dfrac{n^2}{n+1}\right\}_{n=1}^{\infty}$ diverges.

(e) Prove that $\left\{2-(-1)^n\right\}_{n=1}^{\infty}$ diverges.

(f) Suppose $\{p_n\}_{n=1}^{\infty}$ is a real sequence such that $\lim_{n\to\infty} p_n < 0$. Prove that $p_n < 0$ for all $n \geq n_0$, for some $n_0 \in \mathbb{N}$.

2.2 Limit Theorems

In this section, we will establish some important properties of real sequences and find limits of some basic sequences that we will encounter in the study of analysis.

2.2T1 Theorem: Sum Product and Quotient

Suppose $\{a_n\}$ and $\{b_n\}$ are real sequences with $\lim_{n\to\infty} a_n = a$ and $\lim_{n\to\infty} b_n = b$. Then

(a) $\lim_{n\to\infty}(a_n + b_n) = a + b$.

(b) $\lim_{n\to\infty}(a_n b_n) = ab$.

(c) If $b \neq 0$ and $b_n \neq 0$ for all $n \in \mathbb{N}$, then $\lim_{n\to\infty}(a_n / b_n) = (a / b)$.

Proof:

(a) Let $\varepsilon > 0$ be given. Then there exist $n_1, n_2 \in \mathbb{N}$ such that

$$|a_n - a| < \varepsilon / 2 \text{ for all } n \geq n_1 \text{ and } |b_n - b| < \varepsilon / 2 \text{ for all } n \geq n_2.$$

Let $n_3 = \max\{n_1, n_2\}$. Then for all $n \geq n_3$,

$$\begin{aligned} |(a_n + b_n) - (a + b)| &= |(a_n - a) + (b_n - b)| \\ &\leq |a_n - a| + |b_n - b| \\ &< \varepsilon / 2 + \varepsilon / 2 = \varepsilon. \end{aligned}$$

Therefore, $\lim_{n\to\infty}(a_n + b_n) = a + b$.

(b) Since $\{a_n\}$ converges, it is bounded, say by M_1, and let $M = \max\{M_1, |b|\}$. Let $\varepsilon > 0$ be given. Since both sequences converge, there exist $n_1, n_2 \in \mathbb{N}$ such that

$$|a_n - a| < \frac{\varepsilon}{2M} \text{ for all } n \geq n_1 \text{ and } |b_n - b| < \frac{\varepsilon}{2M} \text{ for all } n \geq n_2.$$

Let $n_3 = \max\{n_1, n_2\}$. Then for all $n \geq n_3$,

$$|a_n b_n - ab| = |a_n b_n - a_n b + a_n b - ab|$$

$$\leq |a_n b_n - a_n b| + |a_n b - ab|$$

$$= |a_n||b_n - b| + |b||a_n - a|$$

$$\leq M|b_n - b| + |b||a_n - a|$$

$$< M\frac{\varepsilon}{2M} + M\frac{\varepsilon}{2M} = \varepsilon.$$

(c) We will show that $\lim_{n\to\infty}\frac{1}{b_n} = \frac{1}{b}$. Then

$$\lim_{n\to\infty}\frac{a_n}{b_n} = \lim_{n\to\infty}\left(a_n \cdot \frac{1}{b_n}\right) = \lim_{n\to\infty} a_n \lim_{n\to\infty}\frac{1}{b_n} = a\cdot\frac{1}{b} = \frac{a}{b}\cdot$$

Since $\{b_n\}$ converges and $b \neq 0,$ there exists $n_1 \in \mathbb{N}$ such that

$$\big||b_n| - |b|\big| \leq |b_n - b| < |b|/2 \text{ for all } n \geq n_1.$$

Then for all $n \geq n_1,$

$$|b|/2 \leq |b_n| < 3|b|/2.$$

Let $\varepsilon > 0$ be given. Then there exists $n_2 \in \mathbb{N}$ such that

$$|b_n - b| < \frac{|b|^2\varepsilon}{2} \text{ for all } n \geq n_2.$$

Now let $n_3 = \max\{n_1, n_2\}$. Then for all $n \geq n_3,$

$$\left|\frac{1}{b_n} - \frac{1}{b}\right| = \frac{|b_n - b|}{|b_n||b|} \leq \frac{2}{|b|^2}|b_n - b|$$

$$< \frac{2}{|b|^2}\frac{|b|^2\varepsilon}{2} = \varepsilon.$$

This completes the proof.

2.2T1C Corollary:

Suppose $\{a_n\}$ is a convergent real sequences with $\lim_{n\to\infty} a_n = a$ and $c \in \mathbb{R}.$ Then

(a) $\lim_{n\to\infty}(a_n + c) = a + c.$

(b) $\lim_{n\to\infty}(ca_n) = ca.$

Proof: Define $b_n = c$ for all $n \in \mathbb{N}.$ Then $\lim_{n\to\infty} b_n = c.$ (2.1E3(b)) and (2.2T1(a), (b)).

2.2E1 Example: Find $\lim_{n\to\infty}\frac{n-2}{3n-1}$ using the previous theorem and the fact that $\lim_{n\to\infty}\frac{1}{n}=0$ and $\lim_{n\to\infty} c = c.$

$$\lim_{n\to\infty}\frac{n-2}{3n-1}=\lim_{n\to\infty}\frac{1-2/n}{3-1/n}=\frac{\lim_{n\to\infty}(1-2/n)}{\lim_{n\to\infty}(3-1/n)}$$

$$=\frac{\lim_{n\to\infty}1+\lim_{n\to\infty}\left((-2)\cdot\frac{1}{n}\right)}{\lim_{n\to\infty}3+\lim_{n\to\infty}\left((-1)\cdot\frac{1}{n}\right)}=\frac{\lim_{n\to\infty}1+\lim_{n\to\infty}(-2)\cdot\lim_{n\to\infty}\left(\frac{1}{n}\right)}{\lim_{n\to\infty}3+\lim_{n\to\infty}(-1)\cdot\lim_{n\to\infty}\left(\frac{1}{n}\right)}$$

$$=\frac{1+(-2)\cdot 0}{3+(-1)\cdot 0}=\frac{1}{3}.$$

2.2T2 Theorem: Squeeze Theorem

Suppose $\{a_n\}$, $\{b_n\}$, and $\{c_n\}$ are real sequences such that

$$a_n\le c_n\le b_n \text{ for all } n\in\mathbb{N}.$$

If

$$\lim_{n\to\infty}a_n=L=\lim_{n\to\infty}b_n,$$

then the sequence $\{c_n\}$ converges and $\lim_{n\to\infty}c_n=L.$

Proof: Let $\varepsilon>0$ be given. Then there exist $n_1,n_2\in\mathbb{N}$ such that

$$|a_n-L|<\varepsilon \text{ for all } n\ge n_1 \text{ and } |b_n-L|<\varepsilon \text{ for all } n\ge n_2.$$

Let $n_0=\max\{n_1,n_2\}$ and suppose $n\ge n_0$. Then, either $c_n\le L$ or $c_n>L$. If $c_n\le L$, then

$$a_n\le c_n\le L \text{ and } \begin{aligned}|c_n-L|&=|L-c_n|\\&\le|L-a_n|\\&=|a_n-L|<\varepsilon.\end{aligned}$$

If $c_n>L$, then

$$L<c_n\le b_n \text{ and } |c_n-L|\le|b_n-L|<\varepsilon.$$

In either case,

$$|c_n-L|<\varepsilon \text{ whenever } n\ge n_0.$$

Therefore,

$$\lim_{n\to\infty}c_n=L.$$

Note: The squeeze theorem is sometimes used with the following result.

2.2T3 Theorem:

Let $\{a_n\}$ be a real sequence. Then

$$\lim_{n\to\infty} a_n = 0 \text{ if and only if } \lim_{n\to\infty} |a_n| = 0.$$

Proof: Note that $\left||a_n| - 0\right| = |a_n - 0|$ for all $n \in \mathbb{N}$. Now the result follows if we use the definition of the limit of a sequence (2.1D2).

2.2T4 Theorem: Binomial Theorem

Let $n \in \mathbb{N}$ and $a \in \mathbb{R}$. Then

$$(1+a)^n = \sum_{k=0}^{n} \binom{n}{k} a^k = \binom{n}{0} + \binom{n}{1} a + \binom{n}{2} a^2 + \ldots + \binom{n}{n} a^n$$

where

$$\binom{n}{k} = \frac{n!}{(n-k)!k!}$$

with $0! = 1$ and $n! = 1 \cdot 2 \cdots n$ if $n \in \mathbb{N}$.

Proof: Exercise. (Use the principle of mathematical Induction.)

2.2L1 Lemma:

Let $a \in \mathbb{R}$ and $a > 1$.

(a) If $r = \dfrac{m}{n} \in \mathbb{Q}$, define $a^r = \left(a^m\right)^{1/n}$; $a^{1/n}$ has already been defined (1.4T6).

For $n \in \mathbb{N}$, $a^{-n} = \left(a^n\right)^{-1}$. Since $a^0 = 1$, a^r is well defined as long as $\dfrac{m}{n} = \dfrac{p}{q} \in \mathbb{Q} \Rightarrow \left(a^m\right)^{1/n} = \left(a^p\right)^{1/q}$.

Then $a^{r+s} = a^r a^s$ for all $r, s \in \mathbb{Q}$.

(b) If $x \in \mathbb{R}$, let $A(x) = \{a^t \mid t \in \mathbb{Q},\ t \le x\}$. Then

$$a^r = \sup A(r) \text{ if } r \in \mathbb{Q}.$$

(c) Define $a^x = \sup A(x)$ if $x \in \mathbb{R}$. Then

$$a^{x+y} = a^x a^y \text{ for all } x, y \in \mathbb{R}, .$$

Proof:

(a) Suppose $(m/n) = (p/q) \in \mathbb{Q}$. Without loss of generality, we may assume that $p = km$ and $q = kn$ for some $k \in \mathbb{N}$. Then

$$\left(a^p\right)^{1/q}=\left(a^{km}\right)^{1/kn}=\left(\left(a^m\right)^k\right)^{1/kn}=\left(\left(\left(a^m\right)^k\right)^{1/k}\right)^{1/n}=\left(a^m\right)^{1/n}.$$

These steps are justified because $a^{1/n}=y\Leftrightarrow a=y^n$ where $n\in\mathbb{N}$.
Suppose $r=m/n,\ s=p/q\in\mathbb{Q}$. Then

$$a^{r+s}=a^{\frac{m}{n}+\frac{p}{q}}=a^{\frac{mq+np}{nq}}$$
$$=\left(a^{mq+np}\right)^{1/nq}$$
$$=\left(a^{mq}\right)^{1/nq}\left(a^{np}\right)^{1/nq}$$
$$=a^{\frac{mq}{nq}}a^{\frac{np}{nq}}=a^{m/n}a^{p/q}$$
$$=a^ra^s.$$

(b) Let $\alpha=\sup A(r)$. Suppose $a^r\neq\alpha$, i.e., $a^r<\alpha$. Then there exists $t\in\mathbb{Q},\ t<r$ such that

$$a^r<a^t.$$

But it is easily seen that

$$r,t\in\mathbb{Q},\ t<r\Rightarrow a^t<a^r$$

because $a>1$. This is a contradiction. Therefore,

$$a^r=\alpha=\sup A(r).$$

(c) Fix $x,y\in\mathbb{R}$ and let $r,s\in\mathbb{Q}$ be such that

$$r<x \text{ and } s<y.$$

Then

$$r+s<x+y$$

and

$$a^{r+s}=a^ra^s\Rightarrow a^ra^s\leq\sup\left\{a^{r+s}\mid r+s\leq x+y\right\}=a^{x+y}$$
$$\Rightarrow\sup\left\{a^r\mid r\leq x\right\}\sup\left\{a^s\mid s\leq y\right\}\leq a^{x+y}$$
$$\Rightarrow a^xa^y\leq a^{x+y}.$$
$$a^{r+s}=a^ra^s\Rightarrow a^{r+s}\leq\sup\left\{a^r\mid r\leq x\right\}\sup\left\{a^s\mid s\leq y\right\}=a^xa^y$$
$$\Rightarrow\sup\left\{a^{r+s}\mid r+s\leq x+y\right\}\leq a^xa^y$$
$$\Rightarrow a^{x+y}\leq a^xa^y.$$

Hence result.

Note: It was assumed that integral powers of a real number have already been defined but rational powers of a real number were not defined. These have been established in the above lemma along with the definition of $a^x, x \in \mathbb{R}$ with $a > 1$. If $0 < a < 1$, then $a^x = 1/\left(a^{-1}\right)^x$.

2.2T5 Theorem:

(a) If $p > 0$, then $\lim_{n\to\infty} \frac{1}{n^p} = 0$. (The quantity n^p has been defined in the above lemma.)

(b) If $p > 0$, then $\lim_{n\to\infty} \sqrt[n]{p} = 1$.

(c) $\lim_{n\to\infty} \sqrt[n]{n} = 1$.

(d) If $p > 1$ and $\alpha \in \mathbb{R}$, then $\lim_{n\to\infty} \frac{n^\alpha}{p^n} = 0$. ($n^\alpha$ has been defined in the above lemma.)

(e) If $|p| < 1$, then $\lim_{n\to\infty} p^n = 0$.

(f) If $p \in \mathbb{R}$, then $\lim_{n\to\infty} \frac{p^n}{n!} = 0$.

Proof:

(a) Let $p > 0$. Then there exists $r = m/q \in \mathbb{Q}$ such that $m, q \in \mathbb{N}$ and

$$0 < m/q < p.$$

Let $\varepsilon > 0$ be given and choose $n_0 \in \mathbb{N}$ such that

$$n_0 > \left(\varepsilon^{-q}\right)^{1/m}.$$

Then

$$n \ge n_0 \Rightarrow n^m > \epsilon^{-q} \Rightarrow n^{m/q} > 1/\varepsilon.$$

Therefore,

$$\left|\frac{1}{n^p} - 0\right| = \frac{1}{n^p} < \frac{1}{n^r} < \varepsilon \text{ whenever } n \ge n_0.$$

Hence result.

(b) Exercise. Proof is quite similar to that of (c).

(c) For $n \ge 2$, let

$$x_n = \sqrt[n]{n} - 1.$$

Then x_n is positive and

$$n = \left(1 + x_n\right)^n \ge \binom{n}{2} x_n^{\,2} = \frac{n(n-1)}{2} x_n^{\,2}.$$

Therefore,

$$x_n^{\ 2} \le \frac{2}{n-1} \le \frac{4}{n}.$$

This implies that

$$0 \le x_n \le \frac{2}{\sqrt{n}}.$$

Since $\lim_{n\to\infty} \frac{1}{\sqrt{n}} = 0 = \lim_{n\to\infty} 0$, the result follows by (2.2T2).

(d) Let $n_0 \in \mathbb{N}$ be such that $n_0 > \alpha$. Write $p = 1+q$. Then $q > 0$ because $p > 1$. Let $n \in \mathbb{N}$ be such that $n > 2n_0$. Now use the Binomial theorem:

$$\begin{aligned} p^n &= (1+q)^n \\ &> \binom{n}{n_0} q^{n_0} = \frac{n(n-1)\dots(n-n_0+1)}{n_0\,!} q^{n_0}. \end{aligned}$$

Since $n > 2n_0$,

$$n - n_0 + 1 > \frac{n}{2} + 1 > \frac{n}{2}.$$

Therefore,

$$\frac{n(n-1)\dots(n-n_0+1)}{n_0\,!} > \left(\frac{n}{2}\right)^{n_0} \cdot \frac{1}{n_0\,!} = \frac{n^{n_0}}{2^{n_0} n_0\,!}.$$

It follows that

$$0 \le \frac{n^\alpha}{p^n} \le \left(\frac{2^{n_0} n_0\,!}{q^{n_0}}\right) \cdot \frac{1}{n^{n_0 - \alpha}}.$$

Now apply (a) and (2.2T2) to obtain the result.

(e) If $p = 0$, then the result is trivially true. Suppose $0 < |p| < 1$. Let $q = \frac{1}{|p|}$. Then $q > 1$ and

$$\begin{aligned} \lim_{n\to\infty} |p^n| &= \lim_{n\to\infty} |p|^n \\ &= \lim_{n\to\infty} \frac{1}{q^n} = \lim_{n\to\infty} \frac{n^0}{q^n} = 0, \end{aligned}$$

by (d). The result follows by (2.2T3).

(f) Let $n_0 \in \mathbb{N}$ be such that $n_0 > |p|$. Let $n > n_0$. Then

$$\left|\frac{p^n}{n!}\right| = \frac{|p|^n}{n!} < \frac{|p|^n}{n_0\,! n_0^{\,n-n_0}} = \frac{n_0^{\,n_0}}{n_0\,!}\left(\frac{|p|}{n_0}\right)^n.$$

Since $\frac{|p|}{n_0} < 1,$ it follows that $\lim_{n\to\infty}\left|\frac{p^n}{n!}\right| = 0$ by (e) and (2.2T1C(b)). Therefore,

$$\lim_{n\to\infty}\frac{p^n}{n!} = 0.$$

2.2T6 Theorem: Suppose $\{a_n\}$ is a real positive sequence with $\lim_{n\to\infty} a_n = L > 0.$ Then the sequence $\left\{\sqrt[k]{a_n}\right\}$ converges to $\sqrt[k]{L}$, where $k \in \mathbb{N}.$

Proof: If $L = 0,$ then the proof is trivial. If $L \neq 0,$ the case $k = 2$ is left as an exercise with the following hints:

(i) $\left|\sqrt{a_n} - \sqrt{L}\right| = \dfrac{|a_n - L|}{\left|\sqrt{a_n} + \sqrt{L}\right|}$

(ii) There exists $n_1 \in \mathbb{N}$ such that $a_n > \dfrac{L}{2}$ for all $n \geq n_1.$

The general case is deferred to Chapter 3.

2.2E2 Example: Find (a) $\lim_{n\to\infty}\frac{1+(-1)^n}{1+n^{0.5}}$ (b) $\lim_{n\to\infty}\left(\sqrt{n^2+n}-n\right)$ (c) $\lim_{n\to\infty}\frac{2^n+n^3}{3^n+n^2}.$

(a) $$0 \leq \frac{1+(-1)^n}{1+n^{0.5}} \leq \frac{2}{n^{0.5}} \text{ and } \frac{2}{n^{0.5}} \to 0.$$

Therefore, by (2.2T2),

$$\lim_{n\to\infty}\frac{1+(-1)^n}{1+n^{0.5}} = 0.$$

(b) $$0 \leq \sqrt{n^2+n}-n = \frac{\left(\sqrt{n^2+n}-n\right)\left(\sqrt{n^2+n}+n\right)}{\sqrt{n^2+n}+n} = \frac{1}{\sqrt{1+1/n}+1}.$$

Since $1/n \to 0,$

$$\lim_{n\to\infty}\left(\sqrt{n^2+n}-n\right) = \frac{1}{\sqrt{1+0}+1} = \frac{1}{2}. \qquad \text{(2.2T1 and 2.2T6)}$$

(c) $$\frac{2^n+n^3}{3^n+n^2} = \left(\frac{2}{3}\right)^n \frac{1+\frac{n^3}{2^n}}{1+\frac{n^2}{3^n}} \to 0\cdot\frac{1+0}{1+0} = 0. \qquad \text{(2.2T5(d) and (e))}$$

2.2D1 Definition: Infinite Limits

Let $\{a_n\}$ be a real sequence. $\{a_n\}$ is said to **diverge to** ∞ if for every $X > 0,$ there exists $n_0 \in \mathbb{N}$ such that

$$a_n > X \text{ for all } n \geq n_0.$$

In this case we write

$$\lim_{n\to\infty} a_n = \infty, \text{ or } a_n \to \infty \text{ as } n \to \infty, \text{ or simply } a_n \to \infty.$$

$\{a_n\}$ is said to **diverge to** $-\infty$ if for every $X > 0,$ there exists $n_0 \in \mathbb{N}$ such that

$$a_n < -X \text{ for all } n \geq n_0.$$

In this case we write

$$\lim_{n\to\infty} a_n = -\infty, \text{ or } a_n \to -\infty \text{ as } n \to \infty, \text{ or simply } a_n \to -\infty.$$

Note: It can be shown that any sequence that diverges to ∞ or $-\infty$ is unbounded. However, divergence does not imply that the sequence is unbounded. For example, the sequence $\{(-1)^n\}$ is clearly bounded but diverges.

2.2E3 Example: Let $\{a_n\}$ be a real sequence. Prove that if $\lim_{n\to\infty} a_n = \pm\infty,$ then

$$\lim_{n\to\infty} \frac{1}{a_n} = 0.$$

Suppose $\lim_{n\to\infty} a_n = -\infty.$ Let $\varepsilon > 0$ be given. Then there exists $n_0 \in \mathbb{N}$ such that

$$a_n < -\frac{1}{\varepsilon} \text{ for all } n \geq n_0.$$

Therefore,

$$a_n < 0, -a_n > \frac{1}{\varepsilon}, \text{ and } \left|\frac{1}{a_n} - 0\right| = \frac{1}{|a_n|} = \frac{1}{-a_n} < \varepsilon \text{ for all } n \geq n_0$$

$$\Rightarrow \lim_{n\to\infty} \frac{1}{a_n} = 0.$$

Suppose $\lim_{n\to\infty} a_n = \infty.$ Let $\varepsilon > 0$ be given. Then there exists $n_0 \in \mathbb{N}$ such that

$$a_n > \frac{1}{\varepsilon} \text{ for all } n \geq n_0.$$

Therefore,

$$\left|\frac{1}{a_n}-0\right|=\frac{1}{|a_n|}=\frac{1}{a_n}<\varepsilon \text{ for all } n\geq n_0.$$

Thus, $\lim_{n\to\infty}\frac{1}{a_n}=0.$

2.2E4 Example: Let $\{a_n\}$ be a real sequence. Prove that if $a_n>0$ for all $n\in\mathbb{N}$ and $\lim_{n\to\infty}a_n=0$, then $\lim_{n\to\infty}\frac{1}{a_n}=\infty.$

Let $X>0$ be given. Then there exists $n_0\in\mathbb{N}$ such that

$$a_n=|a_n|=|a_n-0|<\frac{1}{X} \text{ for all } n\geq n_0.$$

Therefore,

$$\frac{1}{a_n}>X \text{ for all } n\geq n_0 \text{ and } \lim_{n\to\infty}\frac{1}{a_n}=\infty.$$

2.2E5 Example: Show that the sequence $\left\{\frac{2n+1}{3\sqrt{n}}\right\}$ diverges to infinity.

$$0<\frac{1}{a_n}=\frac{3\sqrt{n}}{2n+1}=\frac{3}{2\sqrt{n}+1/\sqrt{n}}\leq\frac{3}{2\sqrt{n}}.$$

Since $0\to 0$ and $\frac{3}{2\sqrt{n}}\to 0,$

$$\lim_{n\to\infty}\frac{1}{a_n}=0\Rightarrow\lim_{n\to\infty}a_n=\infty.$$

2.2A1 Assignment:

(a) Determine whether the given sequence converges or diverges. If it converges, find the limit and if it diverges, determine whether it has an infinite limit.

(i) $\left\{\frac{2n+1}{3n^2-1}\right\}$ (ii) $\left\{\frac{2n^2+1}{3n-2}\right\}$ (iii) $\left\{\frac{n(-1)^n}{n+1}\right\}$ (iv) $\left\{\frac{n^2 2^n}{n!}\right\}$.

(b) Show that the given sequence converges and find its limit.

(i) $\left\{\frac{1}{\sqrt{n^3+n^2}-n\sqrt{n}}\right\}$ (ii) $\left\{\frac{\sqrt{n}+1}{\sqrt{n^3+n^2}-n\sqrt{n}}\right\}$.

(c) Let $\{a_n\}$ be a real positive sequence. Suppose $\lim_{n\to\infty}\frac{a_{n+1}}{a_n}=L$. Prove the following:

(i) If $L<1$, then $\{a_n\}$ converges to 0.

(ii) If $L > 1$, then $\{a_n\}$ diverges and is unbounded.

(iii) Give examples to show that if $L = 1$, then $\{a_n\}$ may or may not converge.

(d) Use (c) to prove 2.2T5 (d), (e), and (f).

(e) Let $\{a_n\}$ be a real sequence and let $\{\overline{s}_n\}$ be defined as: $\overline{s}_n = \frac{a_1 + a_2 + ... + a_n}{n}$. Suppose $\lim_{n\to\infty} a_n = 0$. Prove that $\lim_{n\to\infty} \overline{s}_n = 0$. Furthermore, deduce that if $\lim_{n\to\infty} a_n = L$, then $\lim_{n\to\infty} \overline{s}_n = L$.

(f) Let $\{a_n\}$ and $\{b_n\}$ be real sequences such that $\lim_{n\to\infty} a_n = \infty$ and $\lim_{n\to\infty} a_n b_n = L \in \mathbb{R}$. Prove that $\lim_{n\to\infty} b_n = 0$.

(g) Suppose $\{a_n\}$ is a real positive sequence such that $\lim_{n\to\infty} \frac{a_n - 1}{a_n + 1}$. Prove that $\lim_{n\to\infty} a_n = 1$.

2.3 Monotonic Sequences

We will consider monotonic sequences of real numbers in this section. We will see that they either converge or diverge to ∞ or $-\infty$.

2.3D1 Definition: Monotonic Sequences

A real sequence $\{a_n\}$ is said to be

(a) **(monotonically) increasing** if $a_n \le a_{n+1}$ for all $n \in \mathbb{N}$,

(b) **(monotonically) decreasing** if $a_n \ge a_{n+1}$ for all $n \in \mathbb{N}$,

(c) **monotonic** if it is either (monotonically) increasing, or (monotonically) increasing.

Note: A strictly monotonic sequence has $\le$ or $\ge$ replaced by $<$ or $>$, respectively.

2.3T1 Theorem:

If $\{a_n\}$ is monotonic and bounded, then it converges.

Proof: Suppose $\{a_n\}$ is (monotonically) increasing. The set $E = \{a_n \mid n \in \mathbb{N}\}$ is nonempty and bounded above and must have a supremum, say L. Let $\varepsilon > 0$ be given. Then there exists $n_0 \in \mathbb{N}$ such that

$$L - \varepsilon < a_{n_0} \le L.$$

Since $\{a_n\}$ is increasing,

$$L - \varepsilon < a_{n_0} \le a_n \le L < L + \varepsilon \text{ for all } n \ge n_0.$$

Thus

$$|a_n - L| < \varepsilon \text{ for all } n \geq n_0.$$

Therefore $\{a_n\}$ converges to L, i.e., $\lim_{n\to\infty} a_n = L$. A similar proof can be given for the case when $\{a_n\}$ is (monotonically) decreasing by considering the infimum of $E = \{a_n \mid n \in \mathbb{N}\}$.

Note: From the above proof, we see that the theorem can be stated as:

(a) If $\{a_n\}$ is increasing and bounded above, then it converges.

(b) If $\{a_n\}$ is decreasing and bounded below, then it converges.

Note: $a_n \leq M$ for all $n \in \mathbb{N} \Rightarrow \lim_{n\to\infty} a_n \leq M$ and $a_n \geq m$ for all $n \in \mathbb{N} \Rightarrow \lim_{n\to\infty} a_n \geq m$.

2.3T1C Corollary: Nested Intervals Property

Let $I_n = [a_n, b_n]$, $n \in \mathbb{N}$ where $a_1 \leq a_2 \leq \ldots \leq a_n \leq \ldots \leq b_n \leq \ldots \leq b_2 \leq b_1$ so that $I_n \supseteq I_{n+1}$ for all $n \in \mathbb{N}$. Then $\bigcap_{n=1}^{\infty} I_n \neq \phi$.

Proof: The sequence $\{a_n\}$ is increasing and bounded above by b_1. Therefore, it converges, to say $a \in \mathbb{R}$. The sequence $\{b_n\}$ is decreasing and bounded below by a_1. Therefore, it converges, to say $b \in \mathbb{R}$. It is quite clear that

$$a_n \leq a \leq b \leq b_n \text{ for all } n \in \mathbb{N}.$$

It follows that $\bigcap_{n=1}^{\infty} I_n = [a, b] \neq \phi$.

2.3E1 Example: Use the results in this section to show that $\lim_{n\to\infty} p^n = 0$ if $|p| < 1$.

Let $p_n = |p^n|$ for all $n \in \mathbb{N}$. Then

$$\begin{aligned} p_{n+1} &= |p^{n+1}| = |p^n||p| \\ &\leq |p^n| = p_n \text{ for all } n \in N. \end{aligned}$$

Furthermore,

$$0 \leq p_n = |p^n| = |p|^n < 1^n = 1 \text{ for all } n \in N.$$

Therefore, by (2.3T1), $\{p_n\}$ converges, say to L. Then

$$\begin{aligned} L &= \lim_{n\to\infty} p_{n+1} \\ &= \lim_{n\to\infty} |p^{n+1}| \\ &= |p| \lim_{n\to\infty} |p^n| \end{aligned}$$

$$= |p| \lim_{n\to\infty} p_n$$
$$= |p| L.$$

This implies that

$$L(1-|p|) = 0.$$

But $|p| \neq 1$. Therefore,

$$\lim_{n\to\infty} |p^n| = \lim_{n\to\infty} p_n$$
$$= L = 0.$$

2.3E2 Example: Euler's Number *e*

Define the sequence $\{a_n\}$ as $a_n = \left(1+\frac{1}{n}\right)^n$ for all $n \in \mathbb{N}$. Then $\lim_{n\to\infty} a_n = e < 3$.

$$a_n = \left(1+\frac{1}{n}\right)^n = \sum_{k=0}^{n} \binom{n}{k} \frac{1}{n^k} = \sum_{k=0}^{n} \frac{n(n-1)\dots(n-k+1)}{k!} \frac{1}{n^k}.$$

The first term on the right hand side is 1, and the $(k+1)$ – term can be written as

$$\frac{n(n-1)\dots(n-k+1)}{k!} \frac{1}{n^k} = \frac{1\left(1-\frac{1}{n}\right)\dots\left(1-\frac{k-1}{n}\right)}{k!} \quad \text{for } k = 1, 2, \dots, n.$$

$$a_{n+1} = \left(1+\frac{1}{n+1}\right)^{n+1}$$
$$= \sum_{k=0}^{n+1} \binom{n+1}{k} \frac{1}{(n+1)^k}$$
$$= \sum_{k=0}^{n} \frac{(n+1)(n)\dots(n-k+2)}{k!} \frac{1}{(n+1)^k}.$$

The first term on the right hand side is 1, and the $(k+1)$ – term can be written as

$$\frac{(n+1)(n)\dots(n+1-k+1)}{k!} \frac{1}{(n+1)^k} = \frac{1\left(1-\frac{1}{n+1}\right)\dots\left(1-\frac{k-1}{n+1}\right)}{k!} \quad \text{for } k = 1, 2, \dots, n.$$

Therefore, $a_n \le a_{n+1}$ for all $n \in \mathbb{N}$. Furthermore, for all $n \in \mathbb{N}$,

$$a_n = 1 + n\cdot\frac{1}{n} + \frac{n(n-1)}{1\cdot 2}\frac{1}{n^2} + \dots + \frac{n(n-1)\dots(n-k+1)}{1\cdot 2\cdot\dots\cdot k}\frac{1}{n^k} + \dots + \frac{n(n-1)\dots 1}{1\cdot 2\cdot\dots\cdot n}\frac{1}{n^n}$$

$$\le 1+1+\frac{1}{1\cdot 2}+\ldots+\frac{1}{1\cdot 2\cdot\ldots\cdot k}+\ldots+\frac{1}{1\cdot 2\cdot\ldots\cdot n}$$

$$\le 1+1+\frac{1}{1\cdot 2}+\ldots+\frac{1}{1\cdot 2\cdot\ldots\cdot 2}+\ldots+\frac{1}{1\cdot 2\cdot\ldots\cdot 2}$$

$$=1+\frac{1}{2^0}+\frac{1}{2^1}+\ldots+\frac{1}{2^{k-1}}+\ldots+\frac{1}{2^{n-1}}$$

$$=1+\frac{1-2^{-n}}{1-2^{-1}}\le 1+\frac{1}{1-2^{-1}}=3.$$

$\{a_n\}$ is increasing and bounded above by 3 and must converge to a limit $e\le 3$. (2.3T1)

Note: The letter e is the standard notation for the above limit of the above sequence.

2.3E3 Example: Find the limit of the following sequences:

(a) $\left\{\left(1+\frac{1}{n}\right)^{2n+1}\right\}$ (b) $\left\{\left(1+\frac{1}{2n}\right)^{3n}\right\}$.

(a) $$\left(1+\frac{1}{n}\right)^{2n+1}=\left(1+\frac{1}{n}\right)^{n}\left(1+\frac{1}{n}\right)^{n}\left(1+\frac{1}{n}\right)\to e\cdot e\cdot(1+0)=e^2.$$

(b) $$\left(1+\frac{1}{2n}\right)^{3n}=\sqrt{\left(\left(1+\frac{1}{m}\right)^{m}\right)^{3}}.$$

Since $m=2n$ and $m\to\infty\Leftrightarrow n\to\infty$, it follows that the limit of the sequence is $\sqrt{e^3}=e^{3/2}$.

2.3E4 Example: The sequence $\{a_n\}$ is defined by:

$$a_{n+1}=\sqrt{2a_n+3} \text{ for all } n\in\mathbb{N} \text{ and } a_1=4.$$

Show that $\{a_n\}$ converges and find its limit.

Since $a_{n+1}=\sqrt{2a_n+3}$ for all $n\in\mathbb{N}$ and $a_1=4$, $a_n>0$ for all $n\in\mathbb{N}$. We will use mathematical induction to prove that $\{a_n\}$ is decreasing. Let $P(n)$: $a_{n+1}\le a_n$ for $n\in\mathbb{N}$.

$$a_2=\sqrt{8+3}\le 4=a_1\Rightarrow P(1) \text{ is true.}$$

Suppose $P(k)$ is true for some $k\in\mathbb{N}$. Then

$$\begin{aligned}(a_{k+2}+a_{k+1})(a_{k+2}-a_{k+1}) &= a_{k+2}^{\ 2}-a_{k+1}^{\ 2}\\ &=(2a_{k+1}+3)-(2a_k+3)\\ &=2(a_{k+1}-a_k)\le 0.\end{aligned}$$

But $a_n > 0$ for all $n \in \mathbb{N}$. Therefore,

$$a_{k+2}-a_{k+1}\le 0,$$

i.e., $P(k+1)$ is true. By mathematical induction, $P(n)$ is true for all $n \in \mathbb{N}$, i.e., $\{a_n\}$ is decreasing. Since $\{a_n\}$ is bounded below by 0, it follows that $\{a_n\}$ converges, say to L (2.3T1). Now take limits of $a_{n+1}=\sqrt{2a_n+3}$, as $n\to\infty$:

$$\lim_{n\to\infty} a_{n+1}=\lim_{n\to\infty}\sqrt{2a_n+3}\Rightarrow L=\sqrt{2L+3}.$$

Therefore,

$$L^2-2L-3=0\Rightarrow L=3 \text{ or } -1.$$

But $a_n > 0$ for all $n \in \mathbb{N}$. Hence, $L\ne -1$ and $L=3$.

2.3E5 Example: The sequence $\{a_n\}$ is defined by:

$$a_{n+1}=3-\frac{2}{a_n} \text{ for all } n\in\mathbb{N} \text{ and } a_1=3.$$

Show that $\{a_n\}$ converges and find its limit.

We first compute a few terms to test the behavior of the sequence: $a_2=\frac{7}{3}$, $a_3=\frac{15}{7}$. The sequence seems to be decreasing and bounded below by 2. We first prove that the sequence is bounded below by 2. Let $P(n)$: $a_n\ge 2$, $n\in\mathbb{N}$. Then

$P(1)$ is true because $a_1=3\ge 2$. Suppose $P(k)$ is true for some $k\in\mathbb{N}$. Then

$$a_{k+1}=3-\frac{2}{a_k}\ge 3-\frac{2}{2}=2.$$

Thus $P(n)$ is true for all $n\in\mathbb{N}$, i.e., $\{a_n\}$ is bounded below by 2. Therefore,

$$\begin{aligned}a_{n+1}-a_n &= 3-\frac{2}{a_n}-a_n\\ &=-\frac{1}{a_n}(a_n-2)(a_n-1)\le 0 \text{ for all } n\in\mathbb{N},\end{aligned}$$

i.e., $\{a_n\}$ is decreasing. Therefore, $\{a_n\}$ converges, say to L (2.3T1). Now take limits:

$$\lim_{n\to\infty} a_{n+1} = \lim_{n\to\infty}\left(3-\frac{2}{a_n}\right) \Rightarrow L = 3-\frac{2}{L}$$

to obtain $L = 3-\frac{2}{L}\cdot$ Therefore,

$$L^2 - 3L + 2 = 0 \Rightarrow L = 1 \text{ or } L = 2.$$

But $a_n \geq 2$ for all $n \in \mathbb{N}$. Hence,

$$L \neq 1 \text{ and } L = 2.$$

2.3A1 Assignment:

(a) The sequence $\{a_n\}$ is defined by: $a_{n+1} = \sqrt{3a_n - 2}$ for all $n \in \mathbb{N}$ and $a_1 = 4$. Show that $\{a_n\}$ converges and find its limit.

(b) The sequence $\{a_n\}$ is defined by: $a_{n+1} = \frac{1}{2}\left(a_n + \frac{4}{a_n}\right)$ for all $n \in \mathbb{N}$ and $a_1 = 3$. Show that $\{a_n\}$ converges and find its limit. (Hint: First show that $a_n \geq 2$ for all $n \in \mathbb{N}$ by using the fact that the geometric mean is no greater than the arithmetic mean.)

2.4 Subsequences and the Bolzano-Weierstrass Theorem

We will consider subsequences and subsequential limits in this section, and establish a version of the Bolzano-Weierstrass theorem that is quite useful in the study of real analysis.

2.4D1 Definition: Subsequence

Let $\{p_n\}$ be a real sequence and consider an increasing sequence $\{n_k\}$ of positive integers. Then the sequence $\{p_{n_k}\}_{k=1}^{\infty}$ is called a **subsequence** of the sequence $\{p_n\}$. If the sequence $\{p_{n_k}\}_{k=1}^{\infty}$ converges to *p*, then *p* is called a **subsequential limit** of the sequence $\{p_n\}$. A point $p \in \mathbb{R}$ is called a **subsequential limit** of the sequence $\{p_n\}$ if there exists a subsequence $\{p_{n_k}\}_{k=1}^{\infty}$ of $\{p_n\}$ that converges to *p*. We also say ∞ (or $-\infty$) is a **subsequential limit** of $\{p_n\}$ if there exists a subsequence $\{p_{n_k}\}_{k=1}^{\infty}$ of $\{p_n\}$ such that $\lim_{k\to\infty} p_{n_k} = \infty$ (or $\lim_{k\to\infty} p_{n_k} = -\infty$).

2.4E1 Example: Show that the only subsequential limits of $\left\{a_n = \frac{1}{n} + (-1)^n\right\}$ are -1 and 1.

If n is odd, i.e., $n = 2k+1$, then

$$a_n = a_{2k+1} = \frac{1}{2k+1} - 1;$$

this subsequence converges to -1. If n is even, i.e., $n = 2k$, then

$$a_n = a_{2k} = \frac{1}{2k} + 1;$$

this subsequence converges to 1. Let $\{n_k\}$ be an increasing sequence of positive integers. If all but a finite number of the terms in $\{n_k\}$ are odd, then $\{a_{n_k}\}$ clearly converges to -1. If all but a finite number of the terms in $\{n_k\}$ are even, then $\{a_{n_k}\}$ clearly converges to 1. $-\infty$ and ∞ cannot be subsequential limits of $\{a_n\}$ because $\{a_n\}$ is bounded. Suppose $\{a_{n_k}\}$ converges, say to L, and $\{n_k\}$ contains an infinite number of both odd and even integers.

Suppose $L > 0$. Then there exists $n_0 \in \mathbb{N}$ such that

$$|a_{n_k} - L| < L/2 \text{ for all } k \ge n_0.$$

This implies that

$$\frac{L}{2} < \frac{1}{n_k} + (-1)^{n_k} < \frac{3L}{2} \text{ for all } k \ge n_0.$$

Since there are infinitely many odd integers in $\{n_k\}$, we can find $k \ge n_0 + 1$ such that n_k is an odd integer. Using this n_k, we obtain

$$\frac{L}{2} < \frac{1}{n_k} + (-1)^{n_k} \le \frac{1}{2} - 1 = -\frac{1}{2}.$$

This is a contradiction. Thus, L cannot be positive. A similar argument can be used to show that L cannot be negative.

Suppose $L = 0$. Then there exists $n_0 \in \mathbb{N}$ such that

$$|a_{n_k} - 0| < 1/2 \text{ for all } k \ge n_0.$$

This implies that

$$-\frac{1}{2} < \frac{1}{n_k} + (-1)^{n_k} < \frac{1}{2} \text{ for all } k \ge n_0.$$

Since there are infinitely many odd odd integers in $\{n_k\}$, we can find $k \ge n +$ such that n_k is an odd integer. Using this n_k, we obtain

$$-\frac{1}{2} < \frac{1}{n_k} + (-1)^{n_k} \le \frac{1}{2} - 1 = -\frac{1}{2}.$$

This is a contradiction. Thus, $L \neq 0$. Therefore, any subsequence that contains an infinite number of both odd and even integers diverges. Hence, the only subsequential limits are -1 and 1.

2.4T1 Theorem:

Let $\{p_n\}$ be a real sequence that converges to $p \in \mathbb{R}$. If $\{p_{n_k}\}$ is a subsequence of $\{p_n\}$, then it converges to *p*.

Proof: Let $\{p_{n_k}\}$ be a subsequence of $\{p_n\}$, and let $\varepsilon > 0$ be arbitrary. Then there exists $n_0 \in \mathbb{N}$ such that

$$|p_n - p| < \varepsilon \text{ for every } n \geq n_0.$$

Since $\{n_k\}$ is strictly increasing, $n_k \geq n_0$ for all $k \geq n_0$. Therefore,

$$\left|p_{n_k} - p\right| < \varepsilon \text{ for all } k \geq n_0,$$

Hence result.

2.4D2 Definition: Limit Point

Let *E* be a subset of $\mathbb{R}$.

(a) A point $p \in \mathbb{R}$ is said to be a **limit point** (or **accumulation point**) of *E* if for every $\varepsilon > 0,\ N_\varepsilon'(p) \cap E \neq \phi,$ where

$$N_\varepsilon'(p) = \{x \in \mathbb{R} \mid |x - p| < \varepsilon,\ x \neq p\} = (p - \varepsilon, p + \varepsilon) \setminus \{p\},$$

i.e., every $\varepsilon -$neighborhood of *p* contains an element *q* of *E* with $q \neq p$.

(b) A point $p \in E$ is said to be an **isolated point** of *E* if it is not a limit point of *E*.

Notation: Let *E* be a subset of $\mathbb{R}$. The set of all limit points of *E* is denoted by E'.

Note: A limit point of a set need not belong to the set. A point $p \in E$ is an isolated point of *E* if there exists $\varepsilon > 0,$ such that $N_\varepsilon(p) \cap E = \{p\}$.

If *E* is a subset of $\mathbb{R}$, then $E \setminus E' = \{x \mid x \text{ is an isolated point of } E\}$.

2.4E2 Example: It is easily seen that $\mathbb{Q}' = \mathbb{R}$ and $\mathbb{N}' = \mathbb{Z}' = \phi$.

2.4E3 Example:

(a) Let $E = (a,b)$, where $a < b$. Show that $E' = [a,b]$.

Suppose $p \in E$. Let $\varepsilon > 0$ be given and $x = p + \min\{\varepsilon, b - p\}/2$. Then $x \in E,\ x \neq p$ and

$$|x-p| = x-p = \min\{\varepsilon, b-p\}/2 \le \varepsilon/2 < \varepsilon.$$

Thus $x \in N_\varepsilon'(p) \cap E$ and $p \in E'$.

Suppose $p = a$. Let $\varepsilon > 0$ be given and $x = p + \min\{\varepsilon, b-a\}/2$. Then $x \in E$, $x \ne p$ and

$$|x-p| = x-p = \min\{\varepsilon, b-a\}/2 \le \varepsilon/2 < \varepsilon.$$

Thus $x \in N_\varepsilon'(p) \cap E$ and $p \in E'$.

Suppose $p = b$. Let $\varepsilon > 0$ be given and $x = p - \min\{\varepsilon, b-a\}/2$. Then $x \in E$, $x \ne p$ and

$$|x-p| = p-x = \min\{\varepsilon, b-a\}/2 \le \varepsilon/2 < \varepsilon.$$

Thus $x \in N_\varepsilon'(p) \cap E$ and $p \in E'$.

Suppose $p < a$. Let $\varepsilon = a - p > 0$. Then $N_\varepsilon(p) \cap E = \phi$. Thus, $p \notin E'$.

Suppose $p > b$. Let $\varepsilon = p - b > 0$. Then $N_\varepsilon(p) \cap E = \phi$. Thus, $p \notin E'$.

Hence, $E' = [a,b]$.

(b) Let $E = \left\{ \frac{1}{n} \middle| n \in \mathbb{N} \right\}$. Show that $E' = \{0\}$.

Let $p = 0$ and $\varepsilon > 0$ be given. Then by the Archimedian property (1.4T4), there exists a positive integer n such that $0 < \frac{1}{n} < \varepsilon$. Thus $N_\varepsilon'(p) \cap E \ne \phi$ and $p \in E'$.

Suppose $p = \frac{1}{n}$ for some $n \in \mathbb{N}$, i.e., $p \in E$. Let $\varepsilon = \frac{1}{n} - \frac{1}{n+1} > 0$. Then it is easily seen that $N_\varepsilon(p) \cap E = \{p\}$, i.e., p is an isolated point of E.

Suppose $p \in (0,1)$ and $p \in E^c$. Then there exists $n \in \mathbb{N}$ such that

$$\frac{1}{n+1} < p < \frac{1}{n}.$$

Let $\varepsilon = \min\left\{ \frac{1}{n} - p, p - \frac{1}{n+1} \right\}$. Then $N_\varepsilon(p) \cap E = \phi$. Thus, $p \notin E'$.

Suppose $p < 0$. Let $\varepsilon = -p > 0$. Then $N_\varepsilon(p) \cap E = \phi$. Thus, $p \notin E'$.

Suppose $p > 1$. Let $\varepsilon = p - 1 > 0$. Then

$$N_\varepsilon(p) \cap E = \phi.$$

Thus, $p \notin E'$. We have shown that the only limit point of E is 0, which is not in E. Therefore, E consists entirely of isolated points of E.

2.4T2 Theorem:

Let E be a subset of $\mathbb{R}$.

(a) If $p \in E'$, then every neighborhood of p contains infinitely many points of E.

(b) If $p \in E'$, then there exists a sequence $\{p_n\}$ in E with $p_n \neq p$ for all $n \in \mathbb{N}$, such that $\lim_{n\to\infty} p_n = p$.

Proof:

(a) Suppose there exists $\delta > 0$ such that $N'_\delta(p) \cap E = \{q_1, q_2, ..., q_n\}$ for some $n \in \mathbb{N}$. Let $\varepsilon = \min\{|p - q_k| \mid k \in \mathbb{N}\} > 0$. Then $N_\varepsilon'(p) \cap E = \phi$, contradicting $p \in E'$. Hence result.

(b) Let $n \in \mathbb{N}$. Then there exists $p_n \in E$ with $p_n \neq p$ such that $|p_n - p| < \dfrac{1}{n}$. The sequence $\{p_n\}$ clearly converges to p (2.2T1C), (2.2T2, 2.2T3).

2.4T2C Corollary: A finite set has no limit points.

Proof: Follows immediately from (2.4T2(a)).

2.4E4 Example: If $p \in \mathbb{R}$, then there exists a sequence $\{p_n\}$ in $\mathbb{Q}$ such that $\lim_{n\to\infty} p_n = p$. This is quite clear because $\mathbb{Q}' = \mathbb{R}$.

2.4T3 Theorem: Bolzano-Weierstrass Theorem
Every bounded infinite subset of $\mathbb{R}$ has a limit point.

Proof: Let S be a bounded infinite subset of $\mathbb{R}$. Then there exists an interval $I_0 = [a,b]$ with $a < b$ such that $S \subseteq I_0$. Then I_0 contains infinitely many points of S. Divide I_0 into two closed subintervals of equal length $\dfrac{b-a}{2^1}$:

$$\left[a, \frac{a+b}{2}\right] \text{ and } \left[\frac{a+b}{2}, b\right].$$

At least one of these intervals contains infinitely many points of S. Call this interval $I_1 = [a_1, b_1]$. For $n \in \mathbb{N}$, once $I_n = [a_n, b_n]$ of length $\dfrac{b-a}{2^n}$ has been defined, divide it into two closed subintervals of equal length $\dfrac{b-a}{2^{n+1}}$:

$$\left[a_n, \frac{a_n+b_n}{2}\right] \text{ and } \left[\frac{a_n+b_n}{2}, b_n\right].$$

At least one of these intervals contains infinitely many points of S. Call this interval $I_{n+1} = [a_{n+1}, b_{n+1}]$. This process gives us a sequence $\{I_n\}$ of closed and bounded intervals satisfying:

(a) $[a,b] = I_0 \supseteq [a_1,b_1] = I_1 \supseteq [a_2,b_2] = I_2 \supseteq \ldots \supseteq I_n = [a_n,b_n] \supseteq \ldots$

(b) The length of $I_n = \dfrac{b-a}{2^n}$ for each $n \in \mathbb{N}$.

(c) $I_n \cap S$ is infinite for each $n \in \mathbb{N}$.

Therefore, by the Nested Intervals Theorem (2.3T1C), $\bigcap_{n=1}^{\infty} I_n \neq \phi$. Let $x \in \bigcap_{n=1}^{\infty} I_n$. If $\varepsilon > 0$ is given, choose $n \in \mathbb{N}$ so that $\dfrac{b-a}{2^n} < \varepsilon$. Then

$$x \in I_n \text{ and } y \in I_n \Rightarrow |y-x| < \varepsilon$$
$$\Rightarrow I_n \subseteq N_\varepsilon(x).$$

Since $I_n \cap S$ is infinite, there exists $y \in S \cap N_\varepsilon(x)$ with $y \neq x$. Therefore, x is a limit point of S.

2.4T3C Corollary: Sequential Version of the Bolzano-Weierstrass Theorem

Every bounded sequence in $\mathbb{R}$ has a convergent subsequence.

Proof: Let $\{p_n\}$ be a bounded sequence in $\mathbb{R}$, and let $S = \{p_n \mid n \in \mathbb{N}\}$. Then S is bounded. If S is infinite, by (2.4T3) S has a limit point $p \in \mathbb{R}$. Choose $n_1 \in \mathbb{N}$ so that

$$|p_{n_1} - p| < \frac{1}{1}.$$

Having chosen $n_1, n_2, \ldots, n_k \in \mathbb{N}$ with

$$n_1 < n_2 < \ldots < n_k$$

choose $n_{k+1} \in \mathbb{N}$ so that

$$|p_{n_{k+1}} - p| < \frac{1}{k+1}$$

and $n_k < n_{k+1}$. Such an integer exists because every neighborhood of p contains infinitely many points of S. The above construction gives us a subsequence $\{p_{n_k}\}$ of $\{p_n\}$ that converges to p.

2.4T4 Theorem:

Let $\{p_n\}$ be a sequence in $\mathbb{R}$, and let p be a limit point of $S = \{p_n \mid n \in \mathbb{N}\}$. Then there exists a subsequence $\{p_{n_k}\}$ of $\{p_n\}$ that converges to p.

Proof: The same constructive proof used in the above corollary.

2.4E4 Example: Show that if $\{p_n\}$ is a real sequence such that every subsequence of $\{p_n\}$ converges to $p \in \mathbb{R}$, then $\{p_n\}$ converges to *p*.

Suppose $\{p_n\}$ does not converge to *p*. Then there exists $\varepsilon > 0$ such that if $n_0 \in \mathbb{N}$, then

$$|p_n - p| \geq \varepsilon \text{ for some } n \geq n_0.$$

Construct a subsequence of $\{p_n\}$ as follows: Let $n_1 \in \mathbb{N}$ be such that

$$|p_{n_1} - p| \geq \varepsilon \text{ where } n_1 \geq 1.$$

Let $n_2 \in \mathbb{N}$ be such that

$$|p_{n_2} - p| \geq \varepsilon \text{ where } n_2 \geq n_1 + 1.$$

After $n_k \in \mathbb{N}$ has been defined for $k \in \mathbb{N}$, let $n_{k+1} \in \mathbb{N}$ be such that

$$|p_{n_{k+1}} - p| \geq \varepsilon \text{ where } n_{k+1} \geq n_k + 1.$$

This subsequence $\{p_{n_k}\}$ does not converge to *p*, because

$$\lim_{k \to \infty} |p_{n_k} - p| \geq \varepsilon > 0.$$

This is a contradiction. Therefore, $\lim_{n \to \infty} p_n = p$.

2.4A1 Assignment:

(a) Construct a real sequences with the given property. Justify your answer.

(i) The only subsequential limits are: -2, 3, and ∞.

(ii) The set of subsequential limits is $\mathbb{N}$.

(b) Let *A* be a nonempty subset of $\mathbb{R}$ that is bounded below and let $\alpha = \inf A$. If $\alpha \notin A$, then prove that $\alpha \in A'$.

(c) Find the limit points and isolated points of: $A = \left\{ \frac{n+1}{2n} + \frac{(-1)^n}{2} \right\}$. Justify your answer.

2.5 Limit Superior and Limit Inferior

The limit superior and limit inferior of a sequence of real numbers always exist in the extended real number system, unlike the limit of a sequence. These concepts are quite important in our study series of real numbers and power series.

2.5D1 Definition:

Let $\{p_n\}$ be a real sequence and for each $k \in \mathbb{N}$ define $l_k = \inf\{p_n \mid n \in \mathbb{N},\ n \geq k\}$ and $h_k = \sup\{p_n \mid n \in \mathbb{N},\ n \geq k\}$. Then $\{l_k\}$ is increasing and $\{h_k\}$ is decreasing. The **limit**

inferior of $\{p_n\}$, denoted by $\underline{\lim}_{n\to\infty} p_n$, is the limit of the increasing sequence $\{l_k\}$, i.e., $\underline{\lim}_{n\to\infty} p_n = \lim_{k\to\infty} l_k$. The **limit superior** of $\{p_n\}$, denoted by $\overline{\lim}_{n\to\infty} p_n$, is the limit of the decreasing sequence $\{h_k\}$, i.e., $\overline{\lim}_{n\to\infty} p_n = \lim_{k\to\infty} h_k$.

Note: The limit superior and the limit inferior of $\{p_n\}$ will always exist because an increasing sequence will always converge or diverge to ∞, whereas a decreasing sequence will always converge or diverge to $-\infty$. It is easily seen that $\underline{\lim}_{n\to\infty} p_n = \lim_{k\to\infty} l_k \le \lim_{k\to\infty} h_k = \overline{\lim}_{n\to\infty} p_n$.

2.5E1 Example: Find the limit superior and limit inferior of the given sequence.

(a) $\left\{a_n = \frac{1}{n} + (-1)^n\right\}$ (b) $\left\{a_n = n\left(1+(-1)^n\right)\right\}$

(a) This is the example in (2.4E1). We have already shown that it has only two subsequential limits: 1 and -1. We will show that limit superior is 1 and the limit inferior is -1. Let $E_k = \{a_n \mid n \ge k\}$. Then

$$E_k = \begin{cases} \left\{\frac{1}{k} - 1, \frac{1}{k+1} + 1, \frac{1}{k+2} - 1, \ldots\right\} & \text{if } k \text{ is odd} \\ \left\{\frac{1}{k} + 1, \frac{1}{k+1} - 1, \frac{1}{k+2} + 1, \ldots\right\} & \text{if } k \text{ is even.} \end{cases}$$

Therefore, $l_k = \inf E_k = -1$ and

$$h_k = \sup E_k = \begin{cases} \frac{1}{k+1} + 1 & \text{if } k \text{ is odd} \\ \frac{1}{k} + 1 & \text{if } k \text{ is even.} \end{cases}$$

It follows that

$$\underline{\lim}_{n\to\infty} a_n = \lim_{k\to\infty} l_k = -1 \text{ and } \overline{\lim}_{n\to\infty} a_n = \lim_{k\to\infty} h_k = 1.$$

(b) $$E_k = \begin{cases} \{0,\ 2(k+1),\ 0,\ 2(k+3),\ 0, \ldots\} & \text{if } k \text{ is odd} \\ \{2k,\ 0,\ 2(k+2),\ 0, \ldots\} & \text{if } k \text{ is even.} \end{cases}$$

Therefore,

$$l_k = \inf E_k = 0 \Rightarrow \underline{\lim}_{n\to\infty} a_n = \lim_{k\to\infty} l_k = 0$$

and

$$h_k = \sup E_k = \infty \Rightarrow \overline{\lim}_{n\to\infty} a_n = \lim_{k\to\infty} h_k = \infty.$$

2.5T1 Theorem: Let $\{p_n\}$ be a real sequence. Then

(a) $\overline{\lim_{n\to\infty}} p_n = \beta \in \mathbb{R}$ if and only if for all $\varepsilon > 0$,

(i) there exists $n_0 \in \mathbb{N}$ such that $p_n < \beta + \varepsilon$ for all $n \geq n_0$, and

(ii) given $n \in \mathbb{N}$, there exists $k \in \mathbb{N}$ with $k \geq n$ such that $p_k > \beta - \varepsilon$.

(b) $\overline{\lim_{n\to\infty}} p_n = \infty$ if and only if given $M > 0$ and $n \in \mathbb{N}$, there exists $k \in \mathbb{N}$ with $k \geq n$ such that $p_k \geq M$.

(c) $\overline{\lim_{n\to\infty}} p_n = -\infty$ if and only if $\lim_{n\to\infty} p_n = -\infty$.

Proof:

(a) Suppose $\overline{\lim_{n\to\infty}} p_n = \beta = \lim_{k\to\infty} h_k = \lim_{k\to\infty} \sup\{p_n \mid n \geq k\}$. Let $\varepsilon > 0$ be given. Then there exists $n_0 \in \mathbb{N}$ such that

$$h_k < \beta + \varepsilon \text{ for all } k \geq n_0.$$

Since $p_n \leq h_k$ for all $n \geq k$,

$$p_n < \beta + \varepsilon \text{ for all } n \geq n_0.$$

Suppose $n \in \mathbb{N}$ is given. Since $\{h_k\}$ is decreasing,

$$h_k \geq \beta \text{ for all } k \in \mathbb{N} \Rightarrow h_n \geq \beta.$$

Since $\lim_{k\to\infty} h_k = \beta$, there exists $k \in \mathbb{N}$ with $k \geq n$ such that

$$p_k > h_n - \varepsilon \geq \beta - \varepsilon.$$

Conversely, assume that (i) and (ii) hold. Let $\varepsilon > 0$ be given. Then there exists $n_0 \in \mathbb{N}$ such that

$$p_n < \beta + \varepsilon \text{ for all } n \geq n_0.$$

Therefore,

$$h_{n_0} = \sup\{p_n \mid n \geq n_0\} \leq \beta + \varepsilon.$$

Since $\{h_k\}$ is decreasing,

$$h_n \leq h_{n_0} \leq \beta + \varepsilon \text{ for all } n \geq n_0.$$

This implies that

$$\overline{\lim_{n\to\infty}} p_n = \lim_{n\to\infty} h_n \leq \beta + \varepsilon.$$

Since $\varepsilon > 0$ was arbitrary,

$$\overline{\lim_{n\to\infty}} p_n \leq \beta.$$

Suppose $\alpha = \overline{\lim_{n\to\infty}} p_n < \beta$. Choose $\varepsilon > 0$ such that $0 < 2\varepsilon < \beta - \alpha$. Then there exists $n_0 \in \mathbb{N}$ such that

$$p_n < \alpha + \varepsilon < \beta - \varepsilon \text{ for all } n \geq n_0.$$

This contradicts (ii). Hence, $\overline{\lim_{n\to\infty}} p_n = \beta$.

(b) Suppose $\overline{\lim_{n\to\infty}} p_n = \infty = \lim_{k\to\infty} h_k = \lim_{k\to\infty} \sup\{p_n \mid n \geq k\}$ and let $M > 0$ and $n \in \mathbb{N}$ be given. If $p_k < M$ for all $k \geq n$, then $h_k = \sup\{p_m \mid m \geq k\} \leq M$ for all $k \geq n$. This implies that $\lim_{k\to\infty} h_k \neq \infty$. This is a contradiction. Therefore, there exists $k \in \mathbb{N}$ with $k \geq n$ such that $p_k \geq M$.

Conversely, assume that given $M > 0$ and $n \in \mathbb{N}$, there exists $k \in \mathbb{N}$ with $k \geq n$ such that $p_k \geq M$. Then the set $\{p_n \mid n \in \mathbb{N},\ n \geq k\}$ is not bounded above. Thus, $\sup\{p_n \mid n \geq k\} = \infty$ for all $n \in \mathbb{N}$. It follows that $\overline{\lim_{n\to\infty}} p_n = \infty$.

(c) Suppose $\overline{\lim_{n\to\infty}} p_n = -\infty = \lim_{k\to\infty} h_k = \lim_{k\to\infty} \sup\{p_n \mid n \geq k\}$. If $\{p_n \mid n \in \mathbb{N}\}$ is bounded below by α, then

$$h_k = \sup\{p_n \mid n \geq k\} \geq \alpha \text{ for all } k \in \mathbb{N} \Rightarrow \overline{\lim_{n\to\infty}} p_n \geq \alpha.$$

This is a contradiction. Therefore, $\{p_n \mid n \in \mathbb{N}\}$ is not bounded below and $\lim_{n\to\infty} p_n = -\infty$.

Conversely, assume that $\lim_{n\to\infty} p_n = -\infty$. From (b), we see that $\overline{\lim_{n\to\infty}} p_n = \infty$ is not possible. From (a), we see that $\overline{\lim_{n\to\infty}} p_n = \beta \in \mathbb{R}$ is not possible. Therefore, $\overline{\lim_{n\to\infty}} p_n = -\infty$.

2.5T2 Theorem: Let $\{p_n\}$ be a real sequence. Then

(a) $\underline{\lim}_{n\to\infty} p_n = \alpha \in \mathbb{R}$ if and only if for all $\varepsilon > 0$,

- (i) there exists $n_0 \in \mathbb{N}$ such that $p_n > \alpha - \varepsilon$ for all $n \geq n_0$, and
- (ii) given $n \in \mathbb{N}$, there exists $k \in \mathbb{N}$ with $k \geq n$ such that $p_k < \alpha + \varepsilon$.

(b) $\underline{\lim}_{n\to\infty} p_n = -\infty$ if and only if given $M > 0$ and $n \in \mathbb{N}$, there exists $k \in \mathbb{N}$ with $k \geq n$ such that $p_k \leq -M$.

(c) $\underline{\lim}_{n\to\infty} p_n = \infty$ if and only if $\lim_{n\to\infty} p_n = \infty$.

Proof: We can use (2.5T1) if we show that $\underline{\lim}_{n\to\infty} p_n = -\overline{\lim_{n\to\infty}}(-p_n)$. Let $q_n = -p_n$ for all $n \in \mathbb{N}$ and let $b_k = \sup\{q_n \mid n \geq k\}$ for all $k \in \mathbb{N}$. Then

$$b_k = \sup\{q_n \mid n \geq k\} = \sup\{-p_n \mid n \geq k\}$$
$$= -\inf\{p_n \mid n \geq k\} = -l_k$$
$$\Rightarrow \lim_{k\to\infty} b_k = -\lim_{k\to\infty} l_k \Rightarrow \overline{\lim_{n\to\infty}}(q_n) = -\underline{\lim_{n\to\infty}}\, p_n$$
$$\Rightarrow \underline{\lim_{n\to\infty}}\, p_n = -\overline{\lim_{n\to\infty}}(-p_n).$$

Now use the previous theorem with $\{q_n\}$ in place of $\{p_n\}$ and $-\alpha$ in place of β.

(a) $$\overline{\lim_{n\to\infty}}(q_n) = \overline{\lim_{n\to\infty}}(-p_n) = -\alpha \in \mathbb{R}$$

if and only if for all $\varepsilon > 0$,

(i) there exists $n_0 \in \mathbb{N}$ such that $q_n = (-p_n) < -\alpha + \varepsilon$ for all $n \geq n_0$

and

(ii) given $n \in \mathbb{N}$, there exists $k \in \mathbb{N}$ with $k \geq n$ such that $q_k = (-p_k) > -\alpha - \varepsilon$.

Rearranging terms, we obtain

$$-\underline{\lim_{n\to\infty}}\, p_n = -\alpha \in \mathbb{R}, \text{ i.e., } \underline{\lim_{n\to\infty}}\, p_n = \alpha \in \mathbb{R}$$

if and only if for all $\varepsilon > 0$,

(i) there exists $n_0 \in \mathbb{N}$ such that $p_n > \alpha - \varepsilon$ for all $n \geq n_0$

and

(ii) given $n \in \mathbb{N}$, there exists $k \in \mathbb{N}$ with $k \geq n$ such that $p_k < \alpha + \varepsilon$.

(b) $$\overline{\lim_{n\to\infty}}(q_n) = \overline{\lim_{n\to\infty}}(-p_n) = \infty$$

if and only if given $M > 0$ and $n \in \mathbb{N}$, there exists $k \in \mathbb{N}$ with $k \geq n$ such that $q_k = (-p_k) \geq M$. Rearranging terms, we obtain

$$-\underline{\lim_{n\to\infty}}\, p_n = \infty, \text{ i.e., } \underline{\lim_{n\to\infty}}\, p_n = -\infty$$

if and only if given $M > 0$ and $n \in \mathbb{N}$, there exists $k \in \mathbb{N}$ with $k \geq n$ such that $p_k \leq -M$.

(c) $$\overline{\lim_{n\to\infty}}(q_n) = \overline{\lim_{n\to\infty}}(-p_n) = -\infty$$

if and only if $\lim_{n\to\infty} q_n = \lim_{n\to\infty}(-p_n) = -\infty$. This gives us

$$-\underline{\lim_{n\to\infty}}\, p_n = -\infty, \text{ i.e., } \underline{\lim_{n\to\infty}}\, p_n = \infty$$

if and only if $-\lim_{n\to\infty} p_n = -\infty$, i.e., $\lim_{n\to\infty} p_n = \infty$.

Note: $\underline{\lim_{n\to\infty}}\, p_n = \infty \Rightarrow \overline{\lim_{n\to\infty}} p_n = \lim_{n\to\infty} p_n = \infty$

and

$$\overline{\lim_{n\to\infty}} p_n = -\infty \Rightarrow \underline{\lim_{n\to\infty}} p_n = \lim_{n\to\infty} p_n = -\infty.$$

2.5T3 Theorem:

Let $\{p_n\}$ be a real sequence. Then

$$\underline{\lim_{n\to\infty}} p_n = \overline{\lim_{n\to\infty}} p_n = (*) \text{ if and only if } \lim_{n\to\infty} p_n = (*)$$

where $(*) = \infty$, or $-\infty$, or $p \in \mathbb{R}$.

Proof: Suppose $\underline{\lim_{n\to\infty}} p_n = \overline{\lim_{n\to\infty}} p_n = p \in \mathbb{R}$. Then given $\varepsilon > 0$ there exist $n_1, n_2 \in \mathbb{N}$ such that

$$p_n < p + \varepsilon \text{ for all } n \geq n_1$$

and

$$p_n > p - \varepsilon \text{ for all } n \geq n_2.$$

(2.5T1(a)) and (2.5T2(a)). Let $n_0 = \min\{n_1, n_2\}$. Then

$$|p_n - p| < \varepsilon \text{ for all } n \geq n_0.$$

Thus,

$$\lim_{n\to\infty} p_n = p.$$

Conversely, suppose $\lim_{n\to\infty} p_n = p$. Then given $\varepsilon > 0$ there exists $n_0 \in \mathbb{N}$ such that

$$p - \varepsilon < p_n < p + \varepsilon \text{ for all } n \geq n_0.$$

Therefore, conditions (a)-(i) and (a)-(ii) of (2.5T1) and (2.5T2) are clearly satisfied. Thus,

$$\underline{\lim_{n\to\infty}} p_n = \overline{\lim_{n\to\infty}} p_n = p.$$

Suppose $\underline{\lim_{n\to\infty}} p_n = \overline{\lim_{n\to\infty}} p_n = \infty$. Then by (2.5T2(c)),

$$\lim_{n\to\infty} p_n = \infty.$$

If $\lim_{n\to\infty} p_n = \infty$, then by (2.5T2(c)),

$$\underline{\lim_{n\to\infty}} p_n = \infty \Rightarrow \overline{\lim_{n\to\infty}} p_n = \infty.$$

Suppose $\underline{\lim_{n\to\infty}} p_n = \overline{\lim_{n\to\infty}} p_n = -\infty$. Then by (2.5T1(c)),

$$\lim_{n\to\infty} p_n = -\infty.$$

If $\lim_{n\to\infty} p_n = -\infty$, then by (2.5T1(c)),

$$\overline{\lim_{n\to\infty}} p_n = -\infty \Rightarrow \underline{\lim_{n\to\infty}} p_n = -\infty.$$

2.5T4 Theorem:

Let $\{a_n\}$ and $\{b_n\}$ be real sequences. Then

$$\underline{\lim}_{n\to\infty} a_n + \underline{\lim}_{n\to\infty} b_n \le \underline{\lim}_{n\to\infty}(a_n+b_n) \le \underline{\lim}_{n\to\infty} a_n + \overline{\lim}_{n\to\infty} b_n \le \overline{\lim}_{n\to\infty}(a_n+b_n) \le \overline{\lim}_{n\to\infty} a_n + \overline{\lim}_{n\to\infty} b_n.$$

Proof: Let $k \in \mathbb{N}$ and $n \ge k$. Then

$$\inf\{a_n \mid n\ge k\} + \inf\{b_n \mid n\ge k\} \le a_n + b_n \text{ for all } n \ge k$$
$$\Rightarrow \inf\{a_n \mid n\ge k\} + \inf\{b_n \mid n\ge k\} \le \inf\{a_n + b_n | n\ge k\}$$
$$\Rightarrow \lim_{k\to\infty}\left[\inf\{a_n \mid n\ge k\} + \inf\{b_n \mid n\ge k\}\right] \le \lim_{k\to\infty} \inf\{a_n + b_n | n\ge k\}$$
$$\Rightarrow \lim_{k\to\infty}\inf\{a_n \mid n\ge k\} + \lim_{k\to\infty}\inf\{b_n \mid n\ge k\} \le \lim_{k\to\infty} \inf\{a_n + b_n | n\ge k\}.$$

This proves that $\underline{\lim}_{n\to\infty} a_n + \underline{\lim}_{n\to\infty} b_n \le \underline{\lim}_{n\to\infty}(a_n+b_n)$. With a similar argument, we can show that

$\overline{\lim}_{n\to\infty}(a_n+b_n) \le \overline{\lim}_{n\to\infty} a_n + \overline{\lim}_{n\to\infty} b_n$.

The middle inequality $\underline{\lim}_{n\to\infty} a_n + \overline{\lim}_{n\to\infty} b_n \le \overline{\lim}_{n\to\infty}(a_n+b_n)$ follows since

$$\inf\{a_n \mid n\ge k\} + b_n \le a_n + b_n \text{ for all } n \ge k$$
$$\Rightarrow \inf\{a_n \mid n\ge k\} + b_n \le \sup\{a_n + b_n | n\ge k\} \text{ for all } n \ge k$$
$$\Rightarrow \inf\{a_n \mid n\ge k\} + \sup\{b_n \mid n\ge k\} \le \sup\{a_n + b_n | n\ge k\}$$
$$\Rightarrow \lim_{k\to\infty}\left[\inf\{a_n \mid n\ge k\} + \sup\{b_n \mid n\ge k\}\right] \le \lim_{k\to\infty} \sup\{a_n + b_n | n\ge k\}$$
$$\Rightarrow \lim_{k\to\infty}\inf\{a_n \mid n\ge k\} + \lim_{k\to\infty}\sup\{b_n \mid n\ge k\} \le \lim_{k\to\infty} \sup\{a_n + b_n | n\ge k\}.$$

2.5T5 Theorem:

Let $\{p_n\}$ be a real sequence and let E be the set of all subsequential limits of $\{p_n\}$. Note that E may contain the limits ∞ and $-\infty$. Then $\underline{\lim}_{n\to\infty} p_n$ and $\overline{\lim}_{n\to\infty} p_n$ are in E and

$\inf E = \underline{\lim}_{n\to\infty} p_n$ and $\sup E = \overline{\lim}_{n\to\infty} p_n$.

Proof: Let $p = \overline{\lim}_{n\to\infty} p_n \in \mathbb{R}$. If $\varepsilon = \frac{1}{1}$, then there exists $n_1 \in \mathbb{N}$ such that

$$p - \frac{1}{1} < p_{n_1} < p + \frac{1}{1}.$$

Such an integer exists by (2.5T2(a)). Suppose we have chosen $n_1, n_2, ..., n_k \in \mathbb{N}$ such that $n_1 < n_2 < ... < n_k$ and

$$p-\frac{1}{j}<p_{n_j}<p+\frac{1}{j} \text{ for } j=1,\ 2,\ldots,\ k.$$

Let $\varepsilon=\frac{1}{k}$ and let $n_{k+1}\in\mathbb{N}$ be the smallest integer greater than n_k such that

$$p-\frac{1}{k+1}<p_{n_{k+1}}<p+\frac{1}{k+1}.$$

Such an integer exists by (2.5T1(a)). Then $\{p_{n_k}\}$ is a subsequence of $\{p_n\}$ that clearly converges to p. Thus, $p\in E$. Since $p\in E$, $p\le\sup E$. Let $\sup E=\alpha$.

If $\alpha=\infty$, then $\{p_n\}$ is not bounded above and this implies that $p=\infty$, a contradiction. Therefore, $\alpha\ne\infty$.

Suppose $\alpha\in\mathbb{R}$ and $\alpha>p$. Then there exists $\beta\in E$ such that

$$p<\beta\le\alpha.$$

Now choose $\varepsilon>0$ such that

$$p<\beta-2\varepsilon, \text{ i.e., } p+\varepsilon<\beta-\varepsilon.$$

Then there exists $n_0\in\mathbb{N}$ such that

$$p_n-p<\varepsilon \text{ for all } n\ge n_0.$$

Therefore,

$$p_n<p+\varepsilon<\beta-\varepsilon \text{ for all } n\ge n_0.$$

If $\{p_{n_k}\}$ is a subsequence of $\{p_n\}$ that converges, then

$$\lim_{k\to\infty}p_{n_k}\le p+\varepsilon<\beta-\varepsilon.$$

This implies that $\beta\notin E$. This is a contradiction. Therefore, $p=\alpha$. If $\alpha=-\infty$, then $p=-\infty$, a contradiction. Therefore, $\alpha\ne-\infty$. Hence, $p=\sup E$.

Suppose $p=\infty$. Then using (2.5T1(b)), we can construct a subsequence $\{p_{n_k}\}$ such that

$$p_{n_k}>k \text{ for all } k\in\mathbb{N}.$$

This implies that $\infty\in E$ and $p=\infty=\sup E$.

Suppose $p=-\infty$. Then by (2.5T1(b)),

$$\lim_{n\to\infty}p_n=-\infty\Rightarrow E=\{-\infty\} \text{ and } p=\sup E.$$

In a similar manner, we can show that $\inf E=\varliminf_{n\to\infty}p_n$.

2.5E2 Example: Find the limit superior and limit inferior of the given sequence using theorem (2.5T5): (a) $\{a_n=1/n+(-1)^n\}$ (b) $\{a_n=n(1+(-1)^n)\}$

(a) By example (2.4E1), we know that $E = \{-1,\ 1\}$. Therefore, $\overline{\lim_{n\to\infty}} a_n = \sup E = 1$ and $\underline{\lim}_{n\to\infty} a_n = \inf E = -1$.

(b) If n is even, then $a_n = 2n$ and if n is odd then $a_n = 0$. This gives us two subsequences $\{a_{2n} = 2n\}$ and $\{a_{2n-1} = 0\}$. The former diverges to ∞ and the latter converges to 0. It is quite clear that these are the only two subsequential limits. Therefore, $E = \{0,\ \infty\}$,

$$\underline{\lim}_{n\to\infty} a_n = \inf E = 0\text{, and } \overline{\lim_{n\to\infty}} a_n = \sup E = \infty.$$

2.5E3 Example: Let $\{a_n\}$ and $\{b_n\}$ be real sequences.

(a) Prove that if $\overline{\lim_{n\to\infty}} |a_n| = 0$, then $\lim_{n\to\infty} a_n = 0$.

(b) Let $b \in \mathbb{R}$. Suppose for every $\varepsilon > 0$, there exists $n_0 \in \mathbb{N}$ such that $b_n > b - \varepsilon$ for all $n \geq n_0$. Prove that $\underline{\lim}_{\to\infty} b \geq b$.

(a) Since $|a_n| \geq 0$ for all $n \in \mathbb{N}$, the set of subsequential limits of $\{|a_n|\}$, say E, could have only nonnegative entries or ∞. Since $\overline{\lim_{n\to\infty}} |a_n| = \sup E$,

$$E = \{0\} \text{ and } \underline{\lim}_{n\to\infty} a_n = 0 = \overline{\lim_{n\to\infty}} |a_n|.$$

It follows that $\lim_{n\to\infty} a_n = 0$. (2.5T3)

(b) Let $l_k = \inf\{b_n \mid n \geq k\}$ for each $k \in \mathbb{N}$. Let $\varepsilon > 0$ be given. Then there exists $n_0 \in \mathbb{N}$ such that

$$b_n > b - \varepsilon \text{ for all } n \geq n_0.$$

This implies that

$$l_k \geq b - \varepsilon \text{ for all } k \geq n_0.$$

Therefore,

$$\underline{\lim}_{n\to\infty} b_n \geq b - \varepsilon.$$

Since $\varepsilon > 0$ was arbitrary, $\underline{\lim}_{n\to\infty} b_n \geq b$.

2.5A1 Assignment:

(a) Find the limit inferior and limit superior of each of the following sequences. Justify your answer.

(i) $\left\{\dfrac{1+(-1)^n n}{1+n}\right\}$ (ii) $\left\{\dfrac{1+(-1)^n}{1+n}\right\}$

(iii) $\left\{2^{(-1)^n}\left(1+\dfrac{1}{n}\right)\right\}$ (iv) $\left\{(-1)^n\left(1-\dfrac{1}{n}\right)+(-1)^{n+1}\left(1+\dfrac{1}{n}\right)\right\}$

(v) $\left\{\left[3/2+(-1)^n\right]^n\right\}.$

(b) Let $\{a_n\}$ and $\{b_n\}$ be real sequences.

(i) Prove that $\overline{\lim_{n\to\infty}}(-a_n) = -\underline{\lim}_{n\to\infty} a_n$.

(ii) Let $b \in \mathbb{R}$. Suppose for every $\varepsilon > 0$, there exists $n_0 \in \mathbb{N}$ such that $b_n < b + \varepsilon$ for all $n \geq n_0$. Prove that $\overline{\lim_{n\to\infty}} b_n \leq b$.

(iii) If $\underline{\lim}_{n\to\infty} a_n = a \in \mathbb{R}$ and $\lim_{n\to\infty} b_n = b > 0$, then show that $\underline{\lim}_{n\to\infty} a_n b_n = ab$.

2.6 Cauchy Sequences

If we know the limit of a sequence, then we may be able to use the definition to prove that the sequence converges. On the other hand if we can prove that a sequence is increasing and bounded above or is decreasing and bounded below, then we know that it converges. In this case, we may be able to determine its limit as well. However, if the sequence is not monotonic, it may not be possible to determine whether it converges or not using the theorems we have established so far. In this section, we consider another criterion for real sequences to determine convergence.

2.6D1 Definition: Cauchy Sequences

A real sequence $\{p_n\}$ is said to be a **Cauchy sequence** if for every $\varepsilon > 0$, there exists $n_0 \in \mathbb{N}$ such that

$$|p_m - p_n| < \varepsilon \text{ for all integers } m, n \geq n_0.$$

Note: In the above definition, the statement "if for every $\varepsilon > 0$, there exists $\in \mathbb{N}$ such that $|p_m - p_n| < \varepsilon$ for all integers $m, n \geq n_0$." is equivalent to the statement "if for every $\varepsilon > 0$, there exists $n_0 \in \mathbb{N}$ such that $|p_{n+k} - p_n| < \varepsilon$ for all $n \geq n_0$ and $k \in \mathbb{N}$." Therefore, if $\{p_n\}$ is a Cauchy sequence in $\mathbb{R}$, then $\lim_{n\to\infty} |p_{n+k} - p_n| = 0$ for every $k \in \mathbb{N}$. However the converse, "If $\lim_{n\to\infty} |p_{n+k} - p_n| = 0$ for every $k \in \mathbb{N}$, then $\{p_n\}$ is a Cauchy sequence.", is not true as shown in the following example.

2.6E1 Example: Define the real sequence $\{s_n\}$ by

$$s_n = \sum_{j=1}^{n} \frac{1}{j} = 1 + \frac{1}{2} + \ldots + \frac{1}{n}.$$

Then for a fixed $k \in \mathbb{N}$,

$$\left|s_{n+k}-s_n\right|=\frac{1}{n+1}+\frac{1}{n+2}+\ldots+\frac{1}{n+k}.$$

The right side of the identity has a finite number of terms and we can take the limit of each term separately and add these limits. This gives

$$\lim_{n\to\infty}\left|s_{n+k}-s_n\right|=0+0+\ldots+0=0.$$

However for any $n\in\mathbb{N}$,

$$\begin{aligned}\left|s_{n+n}-s_n\right|&=\frac{1}{n+1}+\frac{1}{n+2}+\ldots+\frac{1}{n+n}\\&>n\cdot\frac{1}{n+n}=\frac{1}{2}.\end{aligned}$$

This shows that $\{s_n\}$ is not a Cauchy sequence because we have shown that there is no $n_0\in\mathbb{N}$ for which the statement holds when $\varepsilon=\frac{1}{2}$.

2.6T1 Theorem:

Every convergent real sequence is a Cauchy sequence.

Proof: Let $\{p_n\}$ be a real sequence that converges to $p\in\mathbb{R}$ and let $\varepsilon>0$ be arbitrary. Then there exists $n_0\in\mathbb{N}$ such that $|p_n-p|<\varepsilon/2$ for all $n\geq n_0$. Then

$$\begin{aligned}|p_m-p_n|=|p_m-p+p-p_n|&\leq|p_m-p|+|p-p_n|\\&<\varepsilon/2+\varepsilon/2=\varepsilon \text{ for all } m,n\geq n_0.\end{aligned}$$

Hence, $\{p_n\}$ is a Cauchy sequence.

2.6T2 Theorem:

Every real Cauchy sequence is bounded.

Proof: Let $\{p_n\}$ be a real sequence and take $\varepsilon=1$. Then there exists $n_0\in\mathbb{N}$ such that

$$|p_m-p_n|<1$$

for all integers $m,n\geq n_0$. If $m=n_0$, then

$$\left|p_n-p_{n_0}\right|<1 \text{ for all } n\geq n_0.$$

This implies that

$$|p_n|<1+\left|p_{n_0}\right| \text{ for all } n\geq n_0.$$

Let

$$M=\max\left\{|p_1|,\ |p_2|,\ldots,\ \left|p_{n_0-1}\right|,\ 1+\left|p_{n_0}\right|\right\}.$$

Then

$$|p_n|\leq M \text{ for all } n\in\mathbb{N}.$$

This shows that $\{p_n\}$ is bounded.

2.6T3 Theorem:

Suppose $\{p_n\}$ is a real Cauchy sequence that has a convergent subsequence $\{p_{n_k}\}$. Then $\{p_n\}$ converges.

Proof: Suppose $\{p_{n_k}\}$ converges to *p*. Let $\varepsilon > 0$ be given. Then there exists $n_1, n_2 \in \mathbb{N}$ such that

$$|p_{n_k} - p| < \varepsilon / 2 \text{ for all } k \geq n_1$$

and

$$|p_n - p_m| < \varepsilon / 2 \text{ for all } m, n \geq n_2.$$

Let $n_0 = \max\{n_1, n_2\}$. Choose $n_k \in \mathbb{N}$ such that $k \geq n_0$. Then $n_k \geq n_2$ and if $n \geq n_0$, then

$$\begin{aligned} |p_n - p| &\leq |p_n - p_{n_k}| + |p_{n_k} - p| \\ &< \varepsilon / 2 + \varepsilon / 2 = \varepsilon. \end{aligned}$$

Thus, $\{p_n\}$ converges to *p*.

2.6T4 Theorem:

Suppose $\{p_n\}$ is a real sequence. Then $\{p_n\}$ converges if and only if it is a Cauchy sequence.

Proof: We have already shown that if $\{p_n\}$ converges, then it is a Cauchy sequence (2.6T1). Suppose $\{p_n\}$ is a real Cauchy sequence. Then it is bounded (2.6T2) and must have a convergent subsequence (2.4T3C (Bolzano-Weierstrass)). Therefore by (2.6T3), $\{p_n\}$ converges.

Note: The Bolzano-Weierstrass Theorem and the Completeness Axiom (Least Upper Bound Property) are equivalent. They are also equivalent to the following completeness property:

Every Cauchy sequence in $\mathbb{R}$ converges.

2.6E2 Example: Define the real sequence $\{s_n\}$ by

$$s_n = \sum_{j=1}^{n} \frac{1}{j} = 1 + \frac{1}{2} + \ldots + \frac{1}{n}.$$

We have already shown that this sequence is not a Cauchy sequence (2.6E2). Therefore, it diverges. Furthermore, $\{s_n\}$ is clearly strictly increasing. Therefore, it cannot be bounded above. If it is bounded above, it will converge. Therefore, it diverges to ∞. In the study of series, we will see that it is the sequence of partial sums of a divergent series known as the harmonic series.

2.6D1 Definition: Contractive Sequences

A real sequence $\{p_n\}$ is said to be a **contractive sequence** if there exists $b \in \mathbb{R}$ with $0 < b < 1$ such that

$$|p_{n+1}-p_n| \le b|p_n - p_{n-1}| \text{ for all } n \in \mathbb{N},\ n \ge 2.$$

2.6T5 Theorem:

Let $\{p_n\}$ be a real contractive sequence with b as defined above. Then

(a) $\{p_n\}$ converges to some $p \in \mathbb{R}$.

(b) $|p_n - p| \le \dfrac{b}{1-b}|p_n - p_{n-1}|$ for all $n \in \mathbb{N},\ n \ge 2$.

(c) $|p_n - p| \le \dfrac{b^{n-1}}{1-b}|p_2 - p_1|$ for all $n \in \mathbb{N},\ n \ge 2$.

Proof:

(a) Since $\{p_n\}$ is contractive, it follows that

$$|p_{n+1}-p_n| \le b|p_n - p_{n-1}| \le b^2|p_{n-1}-p_{n-2}| \le \ldots \le b^{n-1}|p_2 - p_1|$$

for all $n \in \mathbb{N},\ n \ge 2$. (This can easily be established using the principle of mathematical induction.) Therefore,

$$\begin{aligned}|p_{n+k}-p_n| &\le |p_{n+k}-p_{n+k-1}| + |p_{n+k-1}-p_{n+k-2}| + \ldots + |p_{n+1}-p_n| \\ &\le b^{n+k-2}|p_2-p_1| + b^{n+k-3}|p_2-p_1| + \ldots + b^{n-1}|p_2-p_1| \\ &= \frac{b^{n-1}|p_2-p_1|}{1-b}\left(1-b^k\right) \\ &\le \frac{b^{n-1}|p_2-p_1|}{1-b} \to 0 \text{ as } n \to \infty.\end{aligned}$$

since $0 < b < 1$. The identity

$$\sum_{j=0}^{k-1} b^j = 1 + b + b^2 + \ldots + b^{k-1} = \frac{1-b^k}{1-b} \text{ for all } k \in \mathbb{N} \text{ and } b \ne 1$$

can easily be proved. This implies that for every $\varepsilon > 0$, there exists $n_0 \in \mathbb{N}$ such that

$$|p_{n+k}-p_n| < \varepsilon \text{ for all } n \ge n_0 \text{ and all } k \in \mathbb{N}.$$

Thus, $\{p_n\}$ is a Cauchy sequence and must converge, say to p.

(b) Let $n \in \mathbb{N},\ n \ge 2$ be fixed and let $k \in \mathbb{N}$. Then

$$\begin{aligned}|p_{n+k}-p_n| &\le |p_{n+k}-p_{n+k-1}| + |p_{n+k-1}-p_{n+k-2}| + \ldots + |p_{n+1}-p_n| \\ &\le b^k|p_n-p_{n-1}| + b^{k-1}|p_n-p_{n-1}| + \ldots + b|p_n-p_{n-1}| \\ &= \frac{b\left(1-b^k\right)}{1-b}|p_n-p_{n-1}| \\ &\to \frac{b}{1-b}|p_n-p_{n-1}| \text{ as } k \to \infty.\end{aligned}$$

Therefore,

$$|p - p_n| \le \frac{b}{1-b}|p_n - p_{n-1}| \text{ for all } n \in \mathbb{N},\ n \ge 2.$$

(c) Follows from (b) because

$$|p_n - p_{n-1}| \le b^{n-2}|p_2 - p_1| \text{ for all } n \in \mathbb{N},\ n \ge 2.$$

2.6E3 Example: Suppose the real sequence $\{p_n\}$ satisfies

$$p_{n+1} - p_n = b(p_n - p_{n-1}) \text{ for all } n \in \mathbb{N},\ n \ge 2,$$

where $0 < |b| < 1$. If $p_1 \ne p_2$, then show that $\{p_n\}$ converges and find its limit.

It is quite clear that $\{p_n\}$ is a contractive sequence because

$$|p_{n+1} - p_n| = |b||(p_n - p_{n-1})| \text{ for all } n \in \mathbb{N},\ n \ge 2,$$

where $0 < |b| < 1$. Therefore it converges, say to p. Then

$$\begin{aligned} p_{n+1} - p_1 &= (p_{n+1} - p_n) + (p_n - p_{n-1}) + \ldots + (p_2 - p_1) \\ &= b^{n-1}(p_2 - p_1) + b^{n-2}(p_2 - p_1) + \ldots + (p_2 - p_1) \\ &= \frac{1-b^n}{1-b}(p_2 - p_1). \end{aligned}$$

Therefore, by taking limits, we obtain

$$p - p_1 = \frac{1}{1-b}(p_2 - p_1).$$

It follows that

$$p = \frac{1}{1-b}(p_2 - p_1) + p_1 = \frac{p_2 - bp_1}{1-b}.$$

2.6E4 Example: Suppose the real sequence $\{p_n\}$ satisfies

$$0 < p_n < 1 \text{ and } 3p_{n+1} = p_n^2 + 1 \text{ for all } n \in \mathbb{N}.$$

Prove that $\{p_n\}$ converges and find its limit.

For all $n \in \mathbb{N}$,

$$\begin{aligned} |p_{n+1} - p_n| &= |p_n^2/3 - p_{n-1}^2/3| \\ &= \frac{1}{3}|p_n + p_{n-1}||p_n - p_{n-1}| \\ &\le \frac{2}{3}|p_n - p_{n-1}|. \end{aligned}$$

This shows that $\{p_n\}$ is contractive and must be Cauchy. Therefore, it converges. Furthermore, its limit p, satisfies: $3p = p^2 + 1$ and $p \le 1$. It follows that $p = \frac{3-\sqrt{5}}{2}$. The other root is $\frac{3+\sqrt{5}}{2} > 1$.

2.6A1 Assignment:

(a) Suppose $\{a_n\}$ and $\{b_n\}$ are real Cauchy sequences. Prove directly that $\{a_n + b_n\}$ and $\{a_n b_n\}$ are Cauchy sequences.

(b) For $n \in \mathbb{N}$, define $s_n = 1 + \frac{1}{1!} + \frac{1}{2!} + ... + \frac{1}{n!}$. Prove directly that $\{s_n\}$ is a real Cauchy sequence.

(c) Let $a_1 = 2$ and for $n \in \mathbb{N}$, define $a_{n+1} = \frac{1}{a_n + 1}$. Prove that $\{a_n\}$ is contractive and find $\lim_{n\to\infty} a_n$.

(d) Let $\{a_n\}$ be a real sequence with $a_1 = 1$ and $a_2 = 2$. Suppose $a_{n+2} = ba_{n+1} + (1-b)a_n$ for all $n \in \mathbb{N}$ where $0 < b < 1$. Prove that $\{a_n\}$ is contractive and find $\lim_{n\to\infty} a_n$.

2.7 Series of Real Constants

In this section, we will study the convergence of series of real constants and establish a few well known basic results as well as some results that are usually not encountered in an elementary calculus course.

2.7D1 Definition: Series of Real Constants

Let $\{a_n\}$ be a real sequence and define another real sequence $\{s_n\}$ as follows:

$$s_n = \sum_{k=1}^{n} a_k = a_1 + a_2 + ... + a_n, \quad n \in \mathbb{N}.$$

The sequence $\{s_n\}$ is called a **series** and will be denoted by

$$\sum_{n=1}^{\infty} a_n \quad \text{or} \quad a_1 + a_2 + ... + a_n + ...$$

The term s_n is called the nth partial sum and a_n is called the nth term of the series.

If $\{s_n\}$ converges to s, then we say that the series $\sum_{n=1}^{\infty} a_n$ converges to s, and write

$$\sum_{n=1}^{\infty} a_n = s.$$

The real number s is called the **sum** of the series. If $\{s_n\}$ diverges, then we say that the series $\sum_{n=1}^{\infty} a_n$ diverges.

Note: We may sometimes consider series of the form

$$\sum_{n=0}^{\infty} a_n = a_0 + a_1 + \ldots + a_n + \ldots,$$

or more generally, series of the form

$$\sum_{n=m}^{\infty} a_n = a_m + a_{m+1} + \ldots + a_n + \ldots$$

where m is an integer. For convenience of notation, when there is no possible ambiguity, we may simply use $\sum a_n$.

2.7T1 Theorem: Cauchy Criterion

$\sum a_n$ converges if and only if for every $\varepsilon > 0$, there exists $n_0 \in \mathbb{N}$ such that

$$\left|\sum_{k=m}^{n} a_k\right| < \varepsilon \text{ for all } m, n \in \mathbb{N} \text{ with } n \geq m \geq n_0.$$

Proof: Follows directly from (2.6D1), (2.6T4), and (2.7D1).

2.7T1C Corollary: Test for Divergence

If $\sum a_n$ converges, then $\lim_{n\to\infty} a_n = 0$.

Proof: Let $m = n \geq n_0$ in the statement: "there exists $n_0 \in \mathbb{N}$ such that $\left|\sum_{k=m}^{n} a_k\right| < \varepsilon$ for all $m, n \in \mathbb{N}$ with $n \geq m \geq n_0$." of the previous theorem. Then

$$|a_n| < \varepsilon \text{ for all } n \geq n_0 \Rightarrow \lim_{n\to\infty} a_n = 0.$$

Note: If $\lim_{n\to\infty} a_n \neq 0$ or if $\lim_{n\to\infty} a_n$ does not exist, then the series diverges.

2.7T2 Theorem:

A series with nonnegative terms converges if and only if its partial sums form a bounded sequence.

Proof: Since $a_n \geq 0$, for all $n \in \mathbb{N}$,

$$s_{n+1} - s_n = a_{n+1} \geq 0 \text{ for all } n \in \mathbb{N}.$$

Therefore, $\{s_n\}$ converges if and only if it is bounded (2.3T1). Note that $\{s_n\}$ is bounded below by 0.

2.7T3 Theorem: Geometric Series

Let $a, r \in \mathbb{R}$ with $a \neq 0$. The **geometric series** $\sum_{n=1}^{\infty} ar^{n-1}$ converges if and only if $|r| < 1$.

Proof: If $|r| \geq 1$, then

$$|r|^n \geq 1 \text{ for all } n \geq 0 \Rightarrow \lim_{n\to\infty} |a_n| = \lim_{n\to\infty} |a||r|^n \geq |a|.$$

This implies that $\lim_{\to\infty}$ cannot be 0, if it exists. Therefore, by (2.7T1C), $\sum_{n=0}^{\infty} r^n$ diverges.

Suppose $|r| < 1$. Then

$$\begin{aligned} s_n = \sum_{k=1}^{n} ar^{k-1} &= a + ar + ar^2 + \ldots + ar^n \\ &= \frac{a(1-r^n)}{1-r} \\ &\to \frac{a(1-0)}{1-r} = \frac{a}{1-r} \text{ as } n \to \infty \end{aligned}$$

By (2.2T5(e)). Thus $\sum_{n=1}^{\infty} ar^{n-1}$ converges to $\frac{a}{1-r}$.

Note: a is called the **first term** of the series and r is called the **common ratio**.

2.7T4 Theorem:

Suppose $\{a_n\}$ is a decreasing sequence of nonnegative real numbers. Then the series $\sum_{n=1}^{\infty} a_n$ converges if and only if the series

$$\sum_{k=0}^{\infty} 2^k a_{2^k} = a_1 + 2a_2 + 4a_4 + \ldots$$

converges.

Proof: Since both series contain only nonnegative terms, it suffices to consider boundedness of the series (2.7T2). Let

$$s_n = a_1 + a_2 + \ldots + a_n,\ n \in \mathbb{N} \text{ and } t_k = a_1 + 2a_2 + \ldots + 2^k a_{2^k},\ k \in \mathbb{N}.$$

Then for $n \leq 2^k$,

$$\begin{aligned} s_n &\leq a_1 + (a_2 + a_3) + \ldots + (a_{2^k} + a_{2^k+1} + \ldots + a_{2^{k+1}-1}) \\ &\leq a_1 + 2a_2 + \ldots + 2^k a_{2^k} = t_k. \end{aligned} \tag{a}$$

If $n > 2^k$, then

$$\begin{aligned} 2s_n &\geq 2a_1 + 2a_2 + 2(a_3 + a_4) + \ldots + 2(a_{2^{k-1}+1} + a_{2^{k-1}+2} + \ldots + a_{2^k}) \\ &\geq a_1 + 2a_2 + 4a_4 \ldots + 2^k a_{2^k} = t_k. \end{aligned} \tag{b}$$

By (a) and (b), $\{s_n\}$ and $\{t_k\}$ are both bounded or both unbounded. Therefore, either both series converge or both series diverge.

Note: The above result was established by **Cauchy** and is often referred to as **Cauchy's Condensation Theorem**.

2.7T5 Theorem:

Let $p \in \mathbb{R}$. The *p*-series $\sum_{n=1}^{\infty} \frac{1}{n^p}$ converges if and only if $p > 1$.

Proof: If $p \leq 0$, then

$$\lim_{n\to\infty} a_n = \lim_{n\to\infty} \frac{1}{n^p} \text{ cannot be 0.}$$

Thus, the series diverges (2.7T1C). Suppose $p > 0$. Then

$$2^k a_{2^k} = 2^k 2^{-kp} = \left(2^{(1-p)}\right)^k \text{ and } 2^{(1-p)} < 1 \text{ if and only if } p > 1.$$

Hence, $\sum_{k=1}^{\infty} 2^k a_{2^k}$ converges if and only if $p > 1$ (2.7T3). Therefore by (2.7T4), $\sum_{n=1}^{\infty} \frac{1}{n^p}$ converges if and only if $p > 1$. This completes the proof.

Note: If we take $p = 1$, then we see that the harmonic series diverges. We present this as a corollary to the above theorem.

2.7T5C Corollary:

The harmonic series $\sum_{n=1}^{\infty} \frac{1}{n} = 1 + \frac{1}{2} + \frac{1}{3} + \ldots$ diverges to ∞.

Proof: $p = 1$ shows that the series diverges. Since the series consists of nonnegative terms, it must diverge to ∞.

2.7T6 Theorem:

The series $\sum_{n=0}^{\infty} \frac{1}{n!} = 1 + \frac{1}{1!} + \frac{1}{2!} + \ldots$ converges to $e = \lim_{n\to\infty} \left(1 + \frac{1}{n}\right)^n$.

Proof: $\sum_{n=0}^{\infty} \frac{1}{n!}$ is a series of nonnegative terms and for all $n \in \mathbb{N}$,

$$s_n = \sum_{k=0}^{n} a_k = 1 + 1 + \frac{1}{1\cdot 2} + \frac{1}{1\cdot 2\cdot 3} + \ldots + \frac{1}{1\cdot 2\cdot 3\cdot \ldots \cdot n}$$

$$\leq 1 + 1 + \frac{1}{2} + \frac{1}{2^2} + \ldots + \frac{1}{2^{n-1}}$$

$$\leq 1 + \frac{1}{1 - 1/2} = 3,$$

Therefore, the series converges (2.7T2). Now define

$$t_n = \left(1 + \frac{1}{n}\right)^n \text{ for all } n \in N.$$

Then

$$t_n = 1 + \frac{n}{1!} \cdot \frac{1}{n} + \frac{n(n-1)}{2!} \cdot \frac{1}{n^2} + \frac{n(n-1)(n-2)}{3!} \cdot \frac{1}{n^3} + \ldots + \frac{n(n-1)(n-2)\ldots 2 \cdot 1}{n!} \cdot \frac{1}{n^n}$$

$$= 1 + \frac{1}{1!} + \frac{(1-1/n)}{2!} \cdot + \frac{(1-1/n)(1-2/n)}{3!} + \ldots + \frac{(1-1/n)(1-2/n)\ldots(1-(n-1)/n)}{n!}$$

$$\le 1 + \frac{1}{1!} + \frac{1}{2!} + \ldots + \frac{1}{n!} = s_n \text{ for all } n \in \mathbb{N}.$$

Taking limits as $n \to \infty$, we obtain,

$$\lim_{n\to\infty} t_n \le \lim_{n\to\infty} s_n, \text{ i.e., } e \le \sum_{n=0}^{\infty} \frac{1}{n!}. \qquad \text{(a)}$$

Now fix $m \in N$. Then for $n \in N$, with $n \ge m$,

$$t_n \ge 1 + \frac{n}{1!} \cdot \frac{1}{n} + \frac{n(n-1)}{2!} \cdot \frac{1}{n^2} + \frac{n(n-1)(n-2)}{3!} \cdot \frac{1}{n^3} + \ldots + \frac{n(n-1)(n-m+1)}{m!} \cdot \frac{1}{n^m}$$

$$= 1 + \frac{1}{1!} + \frac{(1-1/n)}{2!} \cdot + \frac{(1-1/n)(1-2/n)}{3!} + \ldots + \frac{(1-1/n)\ldots(1-(m-1)/n)}{m!}.$$

Taking limits as $n \to \infty$, we obtain,

$$e = \lim_{n\to\infty} t_n \ge 1 + \frac{1}{1!} + \frac{1}{2!} + \ldots + \frac{1}{m!} = s_m.$$

Since the above inequality holds for all $m \in N$, we can take limits as $m \to \infty$ to obtain,

$$e \ge \sum_{n=0}^{\infty} \frac{1}{n!}. \qquad \text{(b)}$$

(a) and (b) give us the desired result.

2.7E1 Example: Show that $0 < e - s_n < \frac{1}{n!n}$ for all $n \in \mathbb{N}$, where s_n is the nth partial sum of the series: $\sum_{n=0}^{\infty} \frac{1}{n!} = 1 + \frac{1}{1!} + \frac{1}{2!} + \ldots$

$$e - s_n = \frac{1}{(n+1)!} + \frac{1}{(n+2)!} + \frac{1}{(n+3)!} + \ldots$$

$$< \frac{1}{(n+1)!}\left[1 + \frac{1}{(n+1)} + \frac{1}{(n+1)^2} + \frac{1}{(n+1)^3} + \ldots\right]$$

$$= \frac{1}{(n+1)!}\left[\frac{1}{1 - 1/(n+1)}\right]$$

$$=\frac{1}{(n+1)!}\left[\frac{n+1}{n}\right]=\frac{1}{n!n}.$$

Since the series is of positive real terms, $\{s_n\}$ increases to e and

$$0<e-s_n \text{ for all } n\in\mathbb{N}.$$

Note: The previous example shows us that the series converges fairly rapidly to e. By taking $n=10$, we see that the error in approximating e by s_{10} is less than $\frac{1}{10!10}<2.8\times10^{-8}$. Therefore, we need to sum only the first 11 terms of the series to guarantee an approximation that is accurate to 2.8×10^{-8}.

2.7E2 Example: Suppose e is rational. Then $e=m/n$ for some $m,n\in\mathbb{N}$. By the previous example,

$$0<n!(e-s_n)<\frac{1}{n}.$$

But

$$n!e=(n-1)!ne\in\mathbb{N} \text{ and } n!s_n=n!\left(1+\frac{1}{1!}+\frac{1}{2!}+\ldots+\frac{1}{n!}\right)\in\mathbb{N}.$$

Thus, $n!(e-s_n)$ is an integer that lies strictly between 0 and 1. This is a contradiction. Hence the number e is irrational.

2.7T7 Theorem:

Suppose $\sum a_n$ converges to A, $\sum a_n$ converges to B, and $c\in\mathbb{R}$. Then

(a) $\sum(a_n+b_n)$ converges to $A+B$.

(b) $\sum(ca_n)$ converges to cA.

Proof: Let $A_n=\sum_{k=1}^{n}a_k$ and $B_n=\sum_{k=1}^{n}b_k$ for all $n\in\mathbb{N}$. Then

$$\lim_{n\to\infty}A_n=A \text{ and } \lim_{n\to\infty}B_n=B.$$

(a) By (2.2T1(a)), $\{A_n+B_n\}$ converges to $A+B$. The result follows because for all $n\in\mathbb{N}$,

$$A_n+B_n=\sum_{k=1}^{n}a_k+\sum_{k=1}^{n}b_k=\sum_{k=1}^{n}(a_k+b_k),$$

the nth partial sum of $\sum(a_n+b_n)$.

(b) By (2.2T1C(b)), $\{cA_n\}$ converges to cA. The result follows because for all $n\in\mathbb{N}$,

$$cA_n=c\sum_{k=1}^{n}a_k=\sum_{k=1}^{n}(ca_k),$$

the nth partial sum of $\sum(ca_n)$.

2.7T8 Theorem: Comparison Test

Let $\sum_{n=1}^{\infty} a_n$ and $\sum_{n=1}^{\infty} b_n$ be real series with nonnegative terms. Suppose there exists $n_0 \in \mathbb{N}$ such that $a_n \le b_n$ for all $n \ge n_0$.

(a) If $\sum_{n=1}^{\infty} b_n$ converges, then so does $\sum_{n=1}^{\infty} a_n$.

(b) If $\sum_{n=1}^{\infty} a_n$ diverges, then so does $\sum_{n=1}^{\infty} b_n$.

Proof: We use the Cauchy criterion given in (2.7T1).

(a) Suppose $\sum_{n=1}^{\infty} b_n$ and let $\varepsilon > 0$ be given. Then there exists $n_1 \in \mathbb{N}$ such that

$$\left|\sum_{k=m}^{n} b_k\right| < \varepsilon \text{ for all } m, n \ge n_1.$$

Therefore,

$$\left|\sum_{k=m}^{n} a_k\right| = \sum_{k=m}^{n} a_k \le \sum_{k=m}^{n} b_k = \left|\sum_{k=m}^{n} b_k\right| < \varepsilon \text{ for all } m, n \ge n_1.$$

This completes the proof.

(b) This follows from (a), because the statement in (b) is the contrapositive of (a).

2.7T9 Theorem: Limit Comparison Test

Let $\sum_{n=1}^{\infty} a_n$ and $\sum_{n=1}^{\infty} b_n$ be real series with positive terms and let $\lim_{n\to\infty} \frac{a_n}{b_n} = L \in (0, \infty)$. Then either both series converge or both series diverge.

Proof: The proof of this theorem is left as an exercise (2.7A1(e)). This result is sometimes easier to use than the general comparison test in (2.7T8).

2.7E3 Example: Discuss the convergence of the following series:

(a) $\sum_{n=1}^{\infty} \frac{n-1}{n+1}$ (b) $\sum_{n=1}^{\infty} \frac{2n+1}{n^2+1}$ (c) $\sum_{n=1}^{\infty} \frac{2^n+3}{3^n-1}$.

(a) $$\lim_{n\to\infty} \frac{n-1}{n+1} = \lim_{n\to\infty} \frac{1-1/n}{1+1/n} = \frac{1-0}{1+0} = 1 \ne 0.$$

Therefore, the series diverges (2.7T1C).

(b) $\frac{2n+1}{n^2+1} \sim \frac{2}{n}$ for large n. Therefore, the series should be divergent. We have to show that $a_n > b_n$ for some b_n such that $\sum_{n=1}^{\infty} b_n$ diverges. For all $n \in \mathbb{N}$,

$$\frac{2n+1}{n^2+1} > \frac{2n}{n^2+n^2} = \frac{1}{n}.$$

Since $\sum_{n=1}^{\infty} \frac{1}{n}$ is divergent,

$$\sum_{n=1}^{\infty} \frac{2n+1}{n^2+1} \text{ must be divergent (2.7T8(b)).}$$

(c) $$0 < \frac{2^n+3}{3^n-1} \le \frac{2^n+2\cdot 2^n}{3^n-3^n/2} = \frac{3\cdot 2^n}{3^n/2} = 6\cdot\left(\frac{2}{3}\right)^n \text{ for all } n \in \mathbb{N}.$$

The geometric series $\sum_{n=1}^{\infty} 6\cdot\left(\frac{2}{3}\right)^n$ converges because $|r| = \frac{2}{3} < 1$. Therefore, by (2.7T8(a)), the series converges.

2.7E4 Example: Suppose $\sum_{n=1}^{\infty} a_n$ is a series with real nonnegative terms that converges. Show that the series $\sum_{n=1}^{\infty} \frac{\sqrt{a_n}}{n}$ converges.

The series $\sum_{n=1}^{\infty} \frac{1}{n^2}$ converges (p – series with $p > 1$). Furthermore, because

$$\frac{\sqrt{a_n}}{n} = \sqrt{a_n \cdot \frac{1}{n^2}}$$

$$\le \frac{a_n + 1/n^2}{2}$$

$$= \frac{1}{2}\left[a_n + \frac{1}{n^2}\right],$$

the series $\sum_{n=1}^{\infty} \frac{\sqrt{a_n}}{n}$ converges (2.7T8).

2.7A1 Assignment:

(a) Discuss the convergence of the given series. Justify your answer.

(i) $\sum_{n=1}^{\infty} \left(1+\frac{1}{n}\right)^{-n}$ (ii) $\sum_{n=1}^{\infty} \frac{2n+1}{n^3+1}$ (iii) $\sum_{n=1}^{\infty} \frac{5^n+1}{4^n+1}$

(b) Suppose $\sum_{n=1}^{\infty} a_n$ converges, where $a_n \geq 0$ for all $n \in \mathbb{N}$. Prove that $\sum_{n=1}^{\infty} a_n{}^p$ converges for all $p > 1$.

(c) Suppose $\sum_{n=1}^{\infty} a_n$ converges, where $a_n \geq 0$ for all $n \in \mathbb{N}$. Prove that $\sum_{n=1}^{\infty} \sqrt{\frac{a_n}{n^p}}$ converges if $p > 1$.

(d) Suppose $\sum_{n=1}^{\infty} a_n$ converges, where $a_n \geq 0$ for all $n \in \mathbb{N}$. Prove that $\sum_{n=1}^{\infty} \frac{a_n}{1+a_n}$ converges.

(e) Prove the limit comparison test (2.7T9).

2.8 Further Tests for Convergence and Absolute Convergence

In this section, we develop further tests for convergence of series and introduce the concepts of absolute convergence and conditional convergence.

2.8D1 Definition: Absolute Convergence

A real series $\sum_{n=1}^{\infty} a_n$ is said to be **absolutely convergent** if $\sum_{n=1}^{\infty} |a_n|$ converges.

2.8T1 Theorem:

Suppose $\sum_{n=1}^{\infty} a_n$ converges absolutely. Then $\sum_{n=1}^{\infty} a_n$ converges.

Proof: We use the Cauchy criterion given in (2.7T1). Let $\varepsilon > 0$ be given. Then there exists $n_0 \in \mathbb{N}$ such that

$$\sum_{k=m}^{n} |a_k| < \varepsilon \text{ for all } m, n \geq n_0.$$

Then it follows that

$$\left| \sum_{k=m}^{n} a_k \right| \leq \sum_{k=m}^{n} |a_k| < \varepsilon \text{ for all } m, n \geq n_0.$$

Hence result.

Note: Absolute convergence implies convergence but the convergence does not imply absolute convergence. It will be shown later that $\sum_{n=1}^{\infty} a_n = \sum_{n=1}^{\infty} \frac{(-1)^n}{n} = 1 - \frac{1}{2} + \frac{1}{3} - \ldots$ converges. But the series of absolute values is the divergent harmonic series

$$\sum_{n=1}^{\infty}|a_n| = \sum_{n=1}^{\infty}\frac{1}{n} = 1+\frac{1}{2}+\frac{1}{3}+\ldots$$

2.8L1 Lemma: Summation by Parts

Given two real sequences $\{a_n\}$ and $\{b_n\}$, let $A_n = \sum_{k=1}^{n} a_k$ for all nonnegative integers n; $A_0 = 0$. If the integers p and q satisfy $1 \le p \le q$, then

$$\sum_{n=p}^{q} a_n b_n = \sum_{n=p}^{q-1} A_n\left(b_n - b_{n+1}\right) + A_q b_q - A_{p-1} b_p.$$

Proof:
$$\sum_{n=p}^{q} a_n b_n = \sum_{n=p}^{q}\left(A_n - A_{n-1}\right) b_n$$
$$= \sum_{n=p}^{q} A_n b_n - \sum_{n=p-1}^{q-1} A_n b_{n+1}$$
$$= \sum_{n=p}^{q-1} A_n\left(b_n - b_{n+1}\right) + A_q b_q - A_{p-1} b_p.$$

2.8D1 Definition: Conditional Convergence

A real convergent series $\sum_{n=1}^{\infty} a_n$ is said to be **conditionally convergent** if it is not absolutely convergent.

2.8T2 Theorem: Dirichlet Test for Convergence

Let $\{a_n\}$ and $\{b_n\}$ be real sequences such that

(a) the partial sums A_n of the series $\sum_{n=1}^{\infty} a_n$ form a bounded sequence,

(b) $\{b_n\}$ is decreasing,

(c) $\lim_{n\to\infty} b_n = 0$.

Then $\sum_{n=1}^{\infty} a_n b_n$ converges.

Proof: We use the Cauchy criterion given in (2.7T1) to prove this result. Since the partial sums of the series $\sum_{n=1}^{\infty} a_n$ form a bounded sequence, there exists $M > 0$ such that

$$|A_n| \le M \text{ for all } n \in \mathbb{N}.$$

Let $\varepsilon > 0$ be given. Then there exists $n_0 \in \mathbb{N}$ such that

$$0 \le b_{n_0} < \varepsilon / 2M.$$

If $p,q \in \mathbb{N}$ with $p \ge q \ge n_0$, then

$$\left|\sum_{n=p}^{q} a_n b_n\right| = \left|\sum_{n=p}^{q-1} A_n\left(b_n - b_{n+1}\right) + A_q b_q - A_{p-1} b_p\right|$$

$$\le \sum_{n=p}^{q-1} |A_n|\left(b_n - b_{n+1}\right) + |A_q| b_q + |A_{p-1}| b_p$$

$$\le M\left|\sum_{n=p}^{q-1}\left(b_n - b_{n+1}\right) + b_q + b_p\right|$$

$$= 2Mb_p \le 2Mb_{n_0} < \varepsilon.$$

Hence result.

2.8T3 Theorem: Alternating Series Test

Suppose $\{b_n\}$ is a real decreasing sequence that converges to 0. Then the alternating series $\sum_{n=1}^{\infty}(-1)^{n-1} b_n$ or $\sum_{n=1}^{\infty}(-1)^{n} b_n$ converges.

Proof: Let $a_n = (-1)^{n-1}$ or $a_n = (-1)^n$ for all $n \in \mathbb{N}$. Then the partial sums of the series $\sum_{n=1}^{\infty} a_n$ are clearly bounded by 1, and $\{b_n\}$ satisfies conditions (b) and (c) of (2.8T2). Hence result.

2.8E1 Example: The alternating harmonic series $\sum_{n=1}^{\infty}\frac{(-1)^{n-1}}{n}$ converges. We can apply (2.8T3) because the sequence $\left\{b_n = \frac{1}{n}\right\}$ is clearly decreasing with $\lim_{n\to\infty} b_n = 0.$

2.8T4 Theorem:

Let $\{a_n\}$ be a sequence of positive real numbers. Then

$$\underline{\lim}_{n\to\infty}\frac{a_{n+1}}{a_n} \le \underline{\lim}_{n\to\infty}\sqrt[n]{a_n} \le \overline{\lim}_{n\to\infty}\sqrt[n]{a_n} \le \overline{\lim}_{n\to\infty}\frac{a_{n+1}}{a_n}.$$

Proof: Suppose

$$r = \underline{\lim}_{n\to\infty}\frac{a_{n+1}}{a_n},\ \alpha = \underline{\lim}_{n\to\infty}\sqrt[n]{a_n},\ \beta = \overline{\lim}_{n\to\infty}\sqrt[n]{a_n},\text{ and } R = \overline{\lim}_{n\to\infty}\frac{a_{n+1}}{a_n}.$$

We shall prove the last inequality first. If $R = \infty$, then

$$\beta \le R.$$

Since $\frac{a_{n+1}}{a_n} > 0$ for all $n \in \mathbb{N}$, $R \ge 0$.

Suppose $0 \le R < \infty$. Choose $p > R$. Then there exists $n_0 \in \mathbb{N}$ such that

$$\frac{a_{n+1}}{a_n} \le p \text{ for all } n \ge n_0.$$

In particular,

$$a_{n_0+1} \le pa_{n_0},\ a_{n_0+2} \le pa_{n_0+1}, \ldots,\ a_{n_0+k} \le pa_{n_0+k-1} \text{ for any } k \in \mathbb{N}.$$

Therefore,

$$a_{n_0+k} \le p^k a_{n_0} \text{ for any } k \in \mathbb{N} \Rightarrow a_n \le a_{n_0} p^{-n_0} p^n \text{ for any } n \ge n_0.$$

Hence

$$\sqrt[n]{a_n} \le \left(\sqrt[n]{a_{n_0} p^{-n_0}}\right) p \text{ for any } n \ge n_0 \Rightarrow \beta = \overline{\lim_{n\to\infty}} \sqrt[n]{a_n} \le (1)\, p = p.$$

Since p was an arbitrary number greater than R, we must have $\beta \le R$. The inequality $r \le \alpha$ can be proved using a similar argument. The middle inequality holds because $\underline{\lim}_{n\to\infty} b_n \le \overline{\lim}_{n\to\infty} b_n$ for any real sequence $\{b_n\}$.

Note: If $\lim_{n\to\infty} \left|\frac{a_{n+1}}{a_n}\right|$ exists and equals L, then $\lim_{n\to\infty} \sqrt[n]{|a_n|}$ exists and equals L. But the existence of $\lim_{n\to\infty} \sqrt[n]{|a_n|}$ does not imply the existence $\lim_{n\to\infty} \left|\frac{a_{n+1}}{a_n}\right|$.

2.8T5 Theorem: Root Test

Let $\{a_n\}$ be a real sequence and $\beta = \overline{\lim_{n\to\infty}} \sqrt[n]{|a_n|}$. Then

(a) $\sum a_n$ converges absolutely if $\beta < 1$.

(b) $\sum a_n$ diverges if $\beta > 1$.

(c) The test is inconclusive if $\beta = 1$.

Proof:

(a) Choose $p \in \mathbb{R}$ such that $\beta < p < 1$. Then there exists $n_0 \in \mathbb{N}$ such that

$$\sqrt[n]{|a_n|} < p \text{ for all } n \ge n_0.$$

Therefore,

$$0 \le |a_n| < p^n \text{ for all } n_0 \in \mathbb{N},$$

and $\sum |a_n|$ converges by the comparison test because the geometric series

$$\sum_{n=1}^{\infty} p^n,\ 0 < p < 1$$

converges. Hence result

.

(b) There exists a subsequence $\left\{\sqrt[n_k]{|a_{n_k}|}\right\}$ of $\left\{\sqrt[n]{|a_n|}\right\}$ that converges to β. Then there exists $n_0 \in \mathbb{N}$ such that

$$\left|\sqrt[n_k]{|a_{n_k}|} - \beta\right| < (\beta - 1)/2 \text{ for all } k \geq n_0.$$

This implies that for all $k \geq n_0$,

$$|a_{n_k}| > \left[(\beta+1)/2\right]^n > 1.$$

Therefore, $\lim_{n\to\infty}|a_n| = \lim_{n\to\infty} a_n = 0$ cannot hold. Hence result.

(c) Consider the series $\sum_{n=1}^{\infty}\frac{1}{n^2}$ and $\sum_{n=1}^{\infty}\frac{1}{n}$. The series $\sum_{n=1}^{\infty}\frac{1}{n^2}$ converges and the series $\sum_{n=1}^{\infty}\frac{1}{n}$ diverges. However $\overline{\lim_{n\to\infty}}\sqrt[n]{|a_n|} = 1$ in both cases.

2.8T6 Theorem: Ratio Test

Let $\sum a_n$ be a real series with nonzero terms and let $r = \underline{\lim_{n\to\infty}}\left|\frac{a_{n+1}}{a_n}\right|$ and $R = \overline{\lim_{n\to\infty}}\left|\frac{a_{n+1}}{a_n}\right|$. Then

(a) $\sum a_n$ converges absolutely if $R < 1$.

(b) $\sum a_n$ diverges if $r > 1$.

(c) The test is inconclusive if $r \leq 1 \leq R$.

Proof: Let $\alpha = \underline{\lim_{n\to\infty}}\sqrt[n]{|a_n|}$, and $\beta = \overline{\lim_{n\to\infty}}\sqrt[n]{|a_n|}$. Then by (2.8T4),

$$r \leq \alpha \leq \beta \leq R.$$

(a) $R < 1$ implies that $\beta = \overline{\lim_{n\to\infty}}\sqrt[n]{|a_n|} < 1$. Therefore, by (2.8T5(a)), the series converges absolutely.

(b) $r > 1$ implies that $\beta = \overline{\lim_{n\to\infty}}\sqrt[n]{|a_n|} > 1$. Therefore, by (2.8T5(b)), the series diverges.

(c) Consider the series $\sum_{n=1}^{\infty}\frac{1}{n^2}$ and $\sum_{n=1}^{\infty}\frac{1}{n}$. The series $\sum_{n=1}^{\infty}\frac{1}{n^2}$ converges and the series $\sum_{n=1}^{\infty}\frac{1}{n}$ diverges. However $r = \underline{\lim_{n\to\infty}}\left|\frac{a_{n+1}}{a_n}\right| = 1 = R = \overline{\lim_{n\to\infty}}\left|\frac{a_{n+1}}{a_n}\right|$ in both cases.

2.8E2 Example: Determine whether the series $\sum_{n=1}^{\infty}\frac{(n!)^2}{(2n)!}$ converges or diverges.

$$\left|\frac{a_{n+1}}{a_n}\right| = \frac{\left[(n+1)!\right]^2}{\left[(2n+2)!\right]} \cdot \frac{(2n)!}{(n!)^2}$$
$$= \frac{(n+1)^2}{(2n+2)(2n+1)} = \frac{n+1}{2(2n+1)} \to \frac{1}{4} \text{ as } n \to \infty.$$

Since $\frac{1}{4} < 1$, the series converges (absolutely) by the ratio test (2.8T6).

Note: In the case of series with nonnegative terms, convergence means absolute convergence.

2.8E3 Example: Determine the limits r, R, α, and β as defined in (2.8T4) for the following series:

$$\sum_{n=1}^{\infty} a_n = 1 + 1 + \frac{1}{2} + \frac{1}{3} + \frac{1}{2^2} + \frac{1}{3^2} + \frac{1}{2^3} + \frac{1}{3^3} + \ldots$$

It is easily seen that

$$\left|\frac{a_{n+1}}{a_n}\right| = \begin{cases} \frac{1}{2}\left(\frac{3}{2}\right)^{(n-2)/2} & \text{if } n \text{ is even} \\ \left(\frac{2}{3}\right)^{(n-1)/2} & \text{if } n \text{ is odd.} \end{cases}$$

Therefore, $r = 0$ and $R = \infty$. The ratio test fails in this case. However,

$$\sqrt[n]{|a_n|} = \begin{cases} \sqrt[n]{\left(\frac{1}{3}\right)^{(n-2)/2}} & \text{if } n \text{ is even} \\ \sqrt[n]{\left(\frac{1}{2}\right)^{(n-1)/2}} & \text{if } n \text{ is odd.} \end{cases} = \begin{cases} \frac{1}{\sqrt{3}}\sqrt[n]{3} & \text{if } n \text{ is even} \\ \frac{1}{\sqrt{2}}\sqrt[n]{\sqrt{2}} & \text{if } n \text{ is odd.} \end{cases}$$

It follows that $\alpha = \frac{1}{\sqrt{3}}$ and $\beta = \frac{1}{\sqrt{2}}$. Since $\beta = \frac{1}{\sqrt{2}} < 1$, the series converges (absolutely). It can be shown that the series is the sum of the two geometric series $\sum_{n=1}^{\infty}\left(\frac{1}{2}\right)^{n-1}$ and $\sum_{n=1}^{\infty}\left(\frac{1}{3}\right)^{n-1}$.

These two series converge to $\frac{1}{1-1/2} = 2$ and $\frac{1}{1-1/3} = \frac{3}{2}$, respectively. Therefore, $\sum_{n=1}^{\infty} a_n$ converges to $2 + \frac{3}{2} = \frac{7}{2}$.

2.8E4 Example: Suppose $\sum a_n$ converges and $\{b_n\}$ is monotonic and bounded. Prove that $\sum a_n b_n$ converges.

Suppose $\{b_n\}$ is increasing and bounded above. Then $\{b_n\}$ converges, say to B. Define a sequence $\{c_n\}$ by

$$c_n = B - b_n \text{ for all } n \in \mathbb{N}.$$

Since $B = \sup\{b_n \mid n \in \mathbb{N}\}$, it follows that $\{c_n\}$ is decreasing with limit 0. Since $\sum a_n$ converges, it has bounded partial sums. Therefore, by (2.8T2, Dirichlet Test), $\sum a_n c_n$ converges. Furthermore, $\sum Ba_n$ converges. Therefore, $\sum Ba_n - \sum a_n c_n$ converges. This implies that

$$\sum a_n (B - c_n) = \sum a_n b_n \text{ converges.}$$

If $\{b_n\}$ is decreasing and bounded below, then $\{b_n\}$ converges, say to B. Define a sequence $\{c_n\}$ by

$$c_n = b_n - B \text{ for all } n \in \mathbb{N}.$$

Since $B = \inf\{b_n \mid n \in \mathbb{N}\}$, it follows that $\{c_n\}$ is decreasing with limit 0. Since $\sum a_n$ converges, it has bounded partial sums. Therefore, by (2.8T2, Dirichlet Test),

$$\sum a_n c_n \text{ converges.}$$

Furthermore, $\sum Ba_n$ converges. Therefore,

$$\sum Ba_n + \sum a_n c_n \text{ converges.}$$

This implies that

$$\sum a_n (B + c_n) = \sum a_n b_n \text{ converges.}$$

Note: The hypothesis in the Dirichlet test is weaker than the hypothesis above.

2.8E5 Example: Suppose $\sum a_n$ is absolutely convergent and $\{b_n\}$ is bounded. Show that $\sum a_n b_n$ is absolutely convergent.

There exists $M > 0$ such that

$$|b_n| \le M \text{ for all } n \in \mathbb{N}.$$

Since $\sum a_n$ is absolutely convergent, given $\varepsilon > 0$ there exists $n_0 \in \mathbb{N}$ such that

$$\sum_{k=m}^{n} |a_k| < \frac{\varepsilon}{M} \text{ for all } m, n \ge n_0.$$

Therefore, if $m, n \ge n_0$, then

$$\sum_{k=m}^{n}|a_k b_k| = \sum_{k=m}^{n}|a_k||b_k|$$
$$\le M\sum_{k=m}^{n}|a_k| < M\cdot\frac{\varepsilon}{M} = \varepsilon.$$

Thus, $\sum a_n b_n$ is absolutely convergent.

2.8E6 Example: Test the series $\sum_{n=1}^{\infty}\frac{p^n}{n^p}$ for absolute and conditional convergence, where p is a real number.

The root test is a lot easier to use for this example. If we wanted to use the ratio test, we would have to consider the case $p=0$ separately.

$$\sqrt[n]{|a_n|} = \frac{|p|}{\left(\sqrt[n]{n}\right)^p}$$
$$\to \frac{|p|}{1^p} = |p| \text{ as } n\to\infty.$$

The series converges absolutely for $|p|<1$ and diverges for $|p|>1$. If $p=1$, then

$$\sum_{n=1}^{\infty}\frac{p^n}{n^p} = \sum_{n=1}^{\infty}\frac{1}{n} \text{ diverges.}$$

If $p=-1$, then $\{(-1)^n n\}$ is unbounded and

$$\sum_{n=1}^{\infty}\frac{p^n}{n^p} = \sum_{n=1}^{\infty}(-1)^n n \text{ diverges.}$$

2.8E7 Example: Let $\sum a_n$ be a real series. Prove that $\sum a_n$ converges absolutely if and only if $\sum a_n b_n$ converges for every choice of $b_n \in \{-1,1\}$.

Suppose $\sum a_n$ converges absolutely and $\{b_n\}$ is a sequence with range $\{-1,1\}$. Given $\varepsilon>0$ there exists $n_0\in\mathbb{N}$ such that

$$\sum_{k=m}^{n}|a_k| < \varepsilon \text{ for all } m,n\ge n_0.$$

Therefore, if $m,n\ge n_0$, then

$$\sum_{k=m}^{n}|a_k b_k| = \sum_{k=m}^{n}|a_k||b_k| = \sum_{k=m}^{n}|a_k| < \varepsilon \text{ for all } m,n\ge n_0.$$

Thus, $\sum a_n b_n$ converges (absolutely).

Conversely, assume that $\sum a_n b_n$ converges for every choice of $b_n \in \{-1,1\}$. Define the sequence $\{b_n\}$ as follows:

$$b_n = \begin{cases} 1 & \text{if } a_n \geq 0 \\ -1 & \text{if } a_n < 0. \end{cases}$$

Since $\sum a_n b_n$ converges, given $\varepsilon > 0$ there exists $n_0 \in \mathbb{N}$ such that

$$\left|\sum_{k=m}^{n} a_k b_k\right| < \varepsilon \text{ for all } m,n \geq n_0.$$

But $a_k b_k = |a_k|$ for all $k \in \mathbb{N}$. Therefore,

$$\left|\sum_{k=m}^{n} a_k b_k\right| = \sum_{k=m}^{n} |a_k| < \varepsilon \text{ for all } m,n \geq n_0$$

and $\sum a_n$ converges absolutely.

2.8D3 Definition: Rearrangement of a Series

A real series $\sum a'_n$ is said to be a rearrangement of the series $\sum a_n$ if there exists a one-to-one and onto function $j : \mathbb{N} \to \mathbb{N}$ such that $a'_n = a_{j(n)}$ for all $n \in \mathbb{N}$.

2.8T7 Theorem:

If the real series $\sum a_n$ converges absolutely to A, then every rearrangement of $\sum a_n$ converges to A.

Proof: Let $\sum a'_n$ be a rearrangement of $\sum a_n$ and let s'_n and s_n be their n^{th} partial sums. Given $\varepsilon > 0$ there exists $n_1 \in \mathbb{N}$ such that

$$\sum_{k=m}^{n} |a_k| < \frac{\varepsilon}{2} \text{ for all } m,n \geq n_1.$$

Suppose $a'_n = a_{j(n)}$ for all $n \in \mathbb{N}$, where $j : \mathbb{N} \to \mathbb{N}$ is one-to-one and onto. Choose $n_0 \in \mathbb{N}$ such that $n_0 \geq n_1$ and

$$\{1, 2, \ldots, n_1\} \subseteq \{j(1), j(2), \ldots, j(n_0)\}.$$

Such an integer exists because j is one-to-one and onto. Let $n \geq n_0$ and consider

$$|s_n - s'_n| = \left|\sum_{k=m}^{n} a_k - \sum_{k=m}^{n} a_{j(k)}\right|.$$

The numbers $a_1, a_2, \ldots, a_{n_1}$ will appear in both sums and cancel leaving terms with an index greater than or equal to n_1. Therefore, for all $n \geq n_0$,

$$|s_n - s'_n| = \left|\sum_{k=m}^{n} a_k - \sum_{k=m}^{n} a_{j(k)}\right|$$

$$\leq \sum_{k=m}^{n} |a_k| + \sum_{k=m}^{n} |a_{j(k)}| < \frac{\varepsilon}{2} + \frac{\varepsilon}{2} = \varepsilon.$$

Thus, $\lim_{n\to\infty} s_n = \lim_{n\to\infty} s'_n$, and the rearrangement converges to A.

Note: For conditionally convergent series we have quite an interesting result provided by Riemann.

2.8T8 Theorem:

Let the real series $\sum a_k$ be conditionally convergent and let $\alpha \in \mathbb{R}$. Then there exists a rearrangement $\sum a'_k$ of $\sum a_k$ that converges to α.

Proof: Without loss of generality, we may assume $a_k \neq 0$ for all $k \in \mathbb{N}$. Let $\alpha \in \mathbb{R}$ and define

$$b_k = \frac{1}{2}(|a_k| + a_k),\ c_k = \frac{1}{2}(|a_k| - a_k);\ \ k \in \mathbb{N}.$$

Then

$$b_k - c_k = a_k,\ b_k + c_k = |a_k|;\ k \in \mathbb{N}.$$

Furthermore, if $a_n > 0$, then

$$b_k = a_k \text{ and } c_k = 0;$$

if $a_k < 0$, then

$$b_k = 0 \text{ and } c_k = |a_k|.$$

Since

$$\sum |a_k| = \sum (b_k + c_k),$$

$\sum b_k$ and $\sum c_k$ cannot both converge. Since,

$$\sum a_k = \sum (b_k - c_k),$$

the convergence of $\sum b_k$ implies the convergence of $\sum c_k$ and the convergence of $\sum c_k$ implies the convergence of $\sum b_k$ Therefore, both $\sum b_k$ and $\sum c_k$ diverge. Let

$$p_1,\ p_2,\ p_3, \ldots$$

denote the positive terms of $\sum a_k$ in the original order, and let

$$q_1,\ q_2,\ q_3, \ldots$$

denote the absolute values of the negative terms of $\sum a_k$ in the original order. Then the series $\sum p_k$ and $\sum q_k$ differ from $\sum b_k$ and $\sum c_k$ only by zero terms, and therefore are also divergent. We will now inductively construct sequences $\sum m_k$ and $\sum n_k$ of positive integers such that the rearranged series

$$\sum a_k' = p_1 + \ldots + p_{m_1} - q_1 - \ldots - q_{n_1} + p_{m_1+1} + \ldots + p_{m_2} - q_{n_1+1} - \ldots - q_{n_2} + \ldots$$

has the desired property. Let m_1 be the smallest integer such that

$$x_1 = p_1 + \ldots + p_{m_1} > \alpha.$$

Such an integer exists because $\sum p_k = \infty$. Similarly, let n_1 be the smallest integer such that

$$y_1 = x_1 - q_1 - \ldots - q_{n_1} < \alpha.$$

Suppose $m_1, m_2, \ldots, m_k$ and $n_1, n_2, \ldots, n_k$ have been chosen. Let m_{k+1} be the smallest integer greater than m_k and n_{k+1} be the smallest integer greater than n_k such that

$$x_{k+1} = y_k + p_{m_k+1} + \ldots + p_{m_{k+1}} > \alpha$$

and

$$y_{k+1} = x_{k+1} - q_{n_k+1} - \ldots - q_{n_{k+1}} < \alpha.$$

Such integers exist because $\sum p_k = \infty$ and $\sum q_k = \infty$. Since m_{k+1} and n_{k+1} were chosen to be the smallest integer for which the above inequalities hold,

$$x_{k+1} - p_{m_{k+1}} \le \alpha \Rightarrow 0 < x_{k+1} - \alpha \le p_{m_{k+1}}$$

and

$$y_{k+1} + q_{n_{k+1}} \ge \alpha \Rightarrow 0 < \alpha - y_{k+1} \le q_{n_{k+1}}.$$

Since $\sum a_k$ converges,

$$\lim_{k\to\infty} p_k = \lim_{k\to\infty} q_k = 0.$$

Therefore,

$$\lim_{k\to\infty} x_k = \lim_{k\to\infty} y_k = \alpha.$$

Let

$$s_n' = \sum_{j=1}^{n} a_j' \text{ for all } n \in \mathbb{N}.$$

If the last term of s_n' is a p_j, then there exists a $k \in \mathbb{N}$ such that

$$y_k < s_n' \le x_{k+1}.$$

If the last term of s_n' is a $-q_j$, then there exists a $k \in \mathbb{N}$ such that

$$y_{k+1} < s_n' \le x_{k+1}.$$

In either case, we have

$$\lim_{n\to\infty} s'_n = \alpha.$$

Therefore,

$$\sum a'_k = p_1 + \ldots + p_{m_1} - q_1 - \ldots - q_{n_1} + p_{m_1+1} + \ldots + p_{m_2} - q_{n_1+1} - \ldots - q_{n_2} + \ldots = \alpha.$$

Note: It can be shown that there exist rearrangements $\sum a'_k$ and $\sum a''_k$ of $\sum a_k$ that diverge to $-\infty$ and ∞, respectively.

2.8A1 Assignment:

(a) Suppose both $\sum a_n^2$ and $\sum b_n^2$ converge. Prove that $\sum a_n b_n$ converges absolutely.

(b) Suppose $\{b_n\}$ is decreasing with $\lim_{n\to\infty} b_n = 0$. If $\{b_n\}$ is a real sequence satisfying $|a_n| \le b_n - b_{n+1}$ for all $n \in \mathbb{N}$, prove that $\sum a_n$ converges absolutely.

(c) Prove that if $\sum a_n$ converges absolutely and $\sum a_n$ converges, then $\sum a_n b_n$ converges absolutely.

(d) Give an example to show that the convergence of both $\sum a_n$ and $\sum b_n$ does not imply the convergence of $\sum a_n b_n$.

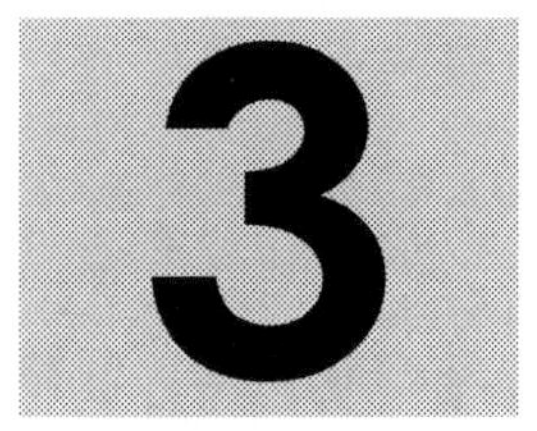

Limits and Continuity

3.1 Structure of Point Sets in $\mathbb{R}$

In this section, we study some of the basic concepts necessary for the study of limits and continuity of real valued functions of a real variable.

3.1D1 Definition: Interior Points and Interior

Let E be a subset of $\mathbb{R}$ and $p \in E$. The point p is said to be an **interior point** of E if there exists $\varepsilon > 0$ such that $N_\varepsilon(p) \subseteq E$. The set of interior points of E, denoted by E°, is called the **interior** of E.

3.1E1 Example: Find the interior of the sets $E = [0,1]$ and $F = [0,1] \cap \mathbb{Q}$.

If $p = 0$, then every neighborhood $N_\varepsilon(p)$ of p contains negative numbers that are not in E. Thus, $p \notin E^\circ$. Suppose $0 < p < 1$. If $\varepsilon = \min\{p, 1-p\}$, then

$$|x - p| < \varepsilon \Rightarrow p - \varepsilon < x < p + \varepsilon.$$

But

$$p - \varepsilon \geq p - p = 0 \text{ and } p + \varepsilon \leq p + 1 - p = 1.$$

Therefore,

$$N_\varepsilon(p) \subseteq E \text{ and } p \in E^\circ.$$

If $p = 1$, then every neighborhood $N_\varepsilon(p)$ of p contains numbers greater than 1 that are not in E. Thus, $p \notin E^\circ$. Hence, $E^\circ = (0,1)$.

Suppose $p \in F = [0,1] \cap \mathbb{Q}$. Then every neighborhood $N_\varepsilon(p)$ of p contains irrational numbers that are not in E. Thus, $p \notin F^\circ$ and $F^\circ = \phi$.

3.1D2 Definition: Open Sets and Closed Sets

(a) A subset E of $\mathbb{R}$ is said to be **open** if $E = E^\circ$.

(b) A subset F of $\mathbb{R}$ is said to be **closed** if its complement $F^c = \mathbb{R} \setminus F$ is open.

Note: From the definition of interior point it is clear that a set E is open if and only if for every $p \in E$ there exists $\varepsilon > 0$ (that depends on p) such that $N_\varepsilon(p) \subseteq E$.

3.1E2 Example: Show that the following sets are open:

(i) $\mathbb{R}$ (ii) $N_\varepsilon(p)$ for any $p \in \mathbb{R}$ and any $\varepsilon > 0$ (iii) E°, where $E \subseteq \mathbb{R}$.

(i) Let $p \in \mathbb{R}$ and $\varepsilon > 0$. Then $N_\varepsilon(p) \subseteq \mathbb{R}$. Hence, $\mathbb{R}$ is open.

(ii) Suppose $q \in N_\varepsilon(p)$. If $\delta = \min\{q - p + \varepsilon, p - q + \varepsilon\}$, then $|x - q| < \delta$ implies that $q - \delta < x < q + \delta$. But $q - \delta \geq q - q + p - \varepsilon = p - \varepsilon$ and $q + \delta \leq q + p - q + \varepsilon = p + \varepsilon$. Therefore, $p - \varepsilon < x < p + \varepsilon$ and $N_\delta(q) \subseteq N_\varepsilon(p)$. Hence, $N_\varepsilon(p)$ is open.

(iii) If $E^\circ = \phi$, then E° is open. Suppose $p \in E^\circ \neq \phi$. Then there exists $\varepsilon > 0$ such that

$$N_\varepsilon(p) \subseteq E.$$

If $q \in N_\varepsilon(p)$, let $\delta = \min\{q - p + \varepsilon, p - q + \varepsilon\}$. Then $|x - q| < \delta$ implies that

$$q - \delta < x < q + \delta.$$

But

$$q - \delta \geq q - q + p - \varepsilon = p - \varepsilon$$

and

$$q + \delta \leq q + p - q + \varepsilon = p + \varepsilon.$$

Therefore,

$$p - \varepsilon < x < p + \varepsilon \text{ and } N_\delta(q) \subseteq N_\varepsilon(p) \subseteq E.$$

Thus, $q \in E^\circ$ and $N_\varepsilon(p) \subseteq E^\circ$. It follows that $(E^\circ)^\circ = E^\circ$ and E° is open.

Note: The empty set ϕ is open because $\phi^\circ = \phi$. Furthermore, since $\phi^c = \mathbb{R}$ and $\mathbb{R}^c = \phi$, both ϕ and $\mathbb{R}$ are closed. Hence, ϕ and $\mathbb{R}$ are both open and closed.

3.1E3 Example: Show that the following sets are not open:

(i) $E = (a, b]$ with $a < b$. (ii) $\mathbb{Q}$ (iii) $\mathbb{N}$.

(i) Every neighborhood $N_\varepsilon(b)$ of b contains points greater than b. Thus $b \notin E^\circ$ and $E^\circ \neq E$. Hence, E is not open.

(ii) Suppose $p \in \mathbb{Q}$. Then every neighborhood $N_\varepsilon(p)$ of p contains irrational numbers that are not in $\mathbb{Q}$. Thus, $p \notin \mathbb{Q}^\circ$ and $\mathbb{Q}^\circ = \phi \neq \mathbb{Q}$. Hence, $\mathbb{Q}$ is not open.

(iii) Suppose $n \in \mathbb{N}$. Then every neighborhood $N_\varepsilon(n)$ of n contains irrational numbers that are not in $\mathbb{N}$. Thus, $p \notin \mathbb{N}$ and $\mathbb{N} = \phi \neq \mathbb{N}$. Hence, $\mathbb{N}$ is not open.

3.1T1 Theorem:

Every open interval in $\mathbb{R}$ is open.

Proof: The only open intervals in $\mathbb{R}$ are of the form $(-\infty,\infty) = \mathbb{R}$, or $(-\infty,b)$, or (a,∞), or (a,b), or ϕ, where $a,b \in \mathbb{R}$ with $a < b$. We already know that $\mathbb{R}$ and ϕ are open.

Consider an interval of the form $(-\infty,b)$. If $p \in (-\infty,b)$, let $\varepsilon = b - p$. Then

$$|x-p| < \varepsilon \Rightarrow x < p + \varepsilon \Rightarrow x < p + b - p = b.$$

Therefore, $N_\varepsilon(p) \subseteq (-\infty,b)$. Thus, $(-\infty,b)$ is open.

Now consider an interval of the form (a,∞). If $p \in (a,\infty)$, let $\varepsilon = p - a$. Then

$$|x-p| < \varepsilon \Rightarrow p - \varepsilon < x \Rightarrow x > p - (p-a) = a.$$

Therefore, $N_\varepsilon(p) \subseteq (a,\infty)$. Hence, (a,∞) is open.

Next consider an interval of the form (a,b). If $p \in (a,b)$, let $\varepsilon = \min\{p-a, b-p\}$. Then

$$|x-p| < \varepsilon \Rightarrow p - \varepsilon < x < p + \varepsilon$$

$$\Rightarrow x > p - (p-a) = a \text{ and } x < p + b - p = b.$$

Thus $N_\varepsilon(p) \subseteq (a,b)$. Hence, (a,b) is open.

3.1T2 Theorem:

(a) If $\{E_\alpha\}_{\alpha \in A}$ is any collection of open subsets of $\mathbb{R}$, then $\bigcup_{\alpha \in A} E_\alpha$ is open.

(b) If $\{E_1, E_2, \ldots, E_n\}$ is any finite collection of open subsets of $\mathbb{R}$, then $\bigcap_{k=1}^{n} E_k$ is open.

Proof:

(a) If $p \in \bigcup_{\alpha \in A} E_\alpha$, then $p \in E_\alpha$ for some $\alpha \in A$. Then $N_\varepsilon(p) \subseteq E_\alpha$ for some $\varepsilon > 0$. This implies that $N_\varepsilon(p) \subseteq \bigcup_{\alpha \in A} E_\alpha$ for some $\varepsilon > 0$. Hence, $\bigcup_{\alpha \in A} E_\alpha$ is open.

(b) If $p \in \bigcap_{k=1}^{n} E_k$, then $p \in E_k$ for all $k = 1,2,\ldots,n$. Since each E_k is open, there exist $\varepsilon_1, \varepsilon_2, \ldots, \varepsilon_n > 0$ such that $N_{\varepsilon_k}(p) \subseteq E_k$ for $k = 1,2,\ldots,n$. Let $\varepsilon = \min\{\varepsilon_1, \varepsilon_2, \ldots, \varepsilon_n\}$. Then $N_\varepsilon(p) \subseteq E_k$ for all $k = 1,2,\ldots,n$. Therefore, $N_\varepsilon(p) \subseteq \bigcap_{k=1}^{n} E_k$. Hence $\bigcap_{k=1}^{n} E_k$ is open.

3.1T3 Theorem:

Every closed interval in $\mathbb{R}$ is closed.

Proof: The only closed intervals in $\mathbb{R}$ are of the form $(-\infty,\infty)=\mathbb{R}$, or $(-\infty,b]$, or $[a,\infty)$, or $[a,b]$, or ϕ, where $a,b\in\mathbb{R}$ with $a\le b$. We already know that $\mathbb{R}$ and ϕ are closed. The complements of $(-\infty,b]$ and $[a,\infty)$ are (b,∞) and $(-\infty,a)$, respectively. Both of these sets are open (3.1T1). Therefore, both $(-\infty,b]$ and $[a,\infty)$ are closed. The complement of $[a,b]$ is $(-\infty,a)\cup(b,\infty)$. Both $(-\infty,a)$ and (b,∞) are open (3.1T1). Therefore, $(-\infty,a)\cup(b,\infty)$ is open (3.1T2). Hence, $[a,b]$ is closed.

3.1T4 Theorem:

(a) If $\{F_\alpha\}_{\alpha\in A}$ is any collection of closed subsets of $\mathbb{R}$, then $\bigcap_{\alpha\in A} F_\alpha$ is closed.

(b) If $\{F_1, F_2,\ldots, F_n\}$ is any finite collection of closed subsets of $\mathbb{R}$, then $\bigcup_{k=1}^{n} F_k$ is closed.

Proof: (a) $\left(\bigcap_{\alpha\in A} F_\alpha\right)^c=\bigcup_{\alpha\in A} F_\alpha^{\ c}$ (DeMorgan's Laws). Each $F_\alpha^{\ c}$ is open because F_α is closed. Therefore, by (3.1T2(a)), $\bigcup_{\alpha\in A} F_\alpha^{\ c}$ is open, i.e., $\left(\bigcap_{\alpha\in A} F_\alpha\right)^c$ is open. Hence, $\bigcap_{\alpha\in A} F_\alpha$ is closed.

(b) $\left(\bigcup_{k=1}^{n} F_k\right)^c=\bigcap_{k=1}^{n} F_k^{\ c}$ (DeMorgan's Laws). Each $F_k^{\ c}$ is open because F_k is closed. Therefore, by (3.1T2(a)), $\bigcap_{k=1}^{n} F_k^{\ c}$ is open, i.e., $\left(\bigcup_{k=1}^{n} F_k\right)^c$ is open. Hence, $\bigcup_{k=1}^{n} F_k$ is closed.

Note: In (3.1T2(b)) and in (3.1T4(b)), an infinite collection of subsets may not give us the required conclusion, e.g., $\bigcap_{n=1}^{\infty}(-1/n,1+1/n)=[0,1]$ is not open and $\bigcup_{n=1}^{\infty}[1/n,1-1/n]=(0,1)$ is not closed.

3.1T5 Theorem:

A subset F of $\mathbb{R}$ is closed if and only if it contains all its limit points, i.e., $F\supseteq F'$.

Proof: Suppose F is closed. Then F^c is open. If $p\in F^c$, then there exists $\varepsilon>0$ such that $N_\varepsilon(p)\subseteq F^c \Rightarrow N_\varepsilon(p)\cap F=\phi$ and $p\notin F'$. It follows that $F\supseteq F'$.

Conversely, suppose $F\supseteq F'$ and let $p\in F^c$. Then $p\notin F'$. This implies that $N_\varepsilon(p)\subseteq F^c$ for some $\varepsilon>0$. Thus, F^c is open and F is closed.

3.1D3 Definition: Closure of a Set

Let E be a subset of $\mathbb{R}$. The **closure** of E, denoted by $\overline{E}$, is defined as $\overline{E} = E \cup E'$, where E' is the set of all limit points of E.

3.1T6 Theorem:

Let E be a subset of $\mathbb{R}$. Then

(a) $\overline{E}$ is closed.

(b) $E = \overline{E}$ if and only if E is closed.

(c) $\overline{E} \subseteq F$ for every closed set $F \subseteq \mathbb{R}$ such that $E \subseteq F$.

Proof:

(a) Let $p \in \overline{E}^c$. Then $p \notin E$ and $p \notin E' \Rightarrow$ there exists $\varepsilon > 0$ such that

$$N_\varepsilon(p) \cap E = \phi.$$

$p \notin E$ and $p \notin E' \Rightarrow$ there exists $\varepsilon > 0$ such that

$$N_\varepsilon(p) \cap E = \phi.$$

Suppose $N_\varepsilon(p) \cap E' \neq \phi$. Let $q \in N_\varepsilon(p) \cap E'$. Choose $\delta > 0$ such that

$$N_\delta(q) \subseteq N_\varepsilon(p).$$

This is possible since $N_\varepsilon(p)$ is open.

$$q \in E',\ N_\delta(q) \cap E \neq \phi \Rightarrow N_\varepsilon(p) \cap E \neq \phi.$$

This is a contradiction. Thus,

$$N_\varepsilon(p) \cap E' = \phi \Rightarrow\ N_\varepsilon(p) \subseteq \overline{E}^c \text{ and } \overline{E}^c \text{ is open.}$$

Hence, $\overline{E}$ is closed.

(b) If $E = \overline{E}$, then E is closed because $\overline{E}$ is closed by (a). If E is closed, then by (3.1T5), $E \supseteq E'$. Hence, $E = \overline{E}$.

(c) If $E \subseteq F$ and F is closed, then $E' \subseteq F' \subseteq \overline{F} = F$ by (b). Thus, $\overline{E} = E \cup E' \subseteq F$.

3.1E4 Example: Show that

(a) $\overline{(A \cup B)} = \overline{A} \cup \overline{B}$

(b) $\overline{(A \cap B)} \subseteq \overline{A} \cap \overline{B}$

(c) Give an example for which $\overline{(A \cap B)} \subset \overline{A} \cap \overline{B}$.

In order to prove (a) and (b), first we need to establish that $(A\cup B)' = A'\cup B'$ and $(A\cap B)' \subseteq A'\cap B'$. If $p\in(A\cup B)'$, then for all $\varepsilon>0$,

$$N_\varepsilon'(p)\cap(A\cup B)\neq\phi,$$

i.e., for all $\varepsilon>0$,

$$N_\varepsilon'(p)\cap A\neq\phi \text{ or } N_\varepsilon'(p)\cap B\neq\phi.$$

Thus, $p\in A'\cup B'$. Suppose $p\in A'\cup B'$. Then for all $\varepsilon>0$,

$$N_\varepsilon'(p)\cap A\neq\phi \text{ or } N_\varepsilon'(p)\cap B\neq\phi.$$

Thus, for all $\varepsilon>0$,

$$N_\varepsilon'(p)\cap(A\cup B)\neq\phi.$$

Therefore, $p\in(A\cup B)'$ Hence, $(A\cup B)' = A'\cup B'$ If $p\in(A\cap B)'$, then for all $\varepsilon>0$,

$$N_\varepsilon'(p)\cap(A\cap B)\neq\phi,$$

i.e., for all $\varepsilon>0$,

$$N_\varepsilon'(p)\cap A\neq\phi \text{ and } N_\varepsilon'(p)\cap B\neq\phi.$$

Thus, $p\in A'\cap B'$ Hence, $(A\cap B)'\subseteq A'\cap B'$.

(a) $$\overline{(A\cup B)} = (A\cup B)\cup(A\cup B)' = A\cup B\cup A'\cup B'$$
$$= A\cup A'\cup B\cup B' = \overline{A}\cup\overline{B}.$$

(b) $$\overline{(A\cap B)} = (A\cap B)\cup(A\cap B)'$$
$$\subseteq (A\cap B)\cup(A'\cap B')$$
$$= (A\cup(A'\cap B'))\cup(B\cup(A'\cap B'))$$
$$\subseteq (A\cup A')\cup(B\cup B') = \overline{A}\cap\overline{B}.$$

(c) Let $A=[0,1]$ and $B=(1,2)$. Then $\overline{(A\cap B)}=\phi\neq\{1\}=\overline{A}\cap\overline{B}$.

3.1T7 Theorem:

Let E be a subset of $\mathbb{R}$ that is bounded. Then $\alpha=\sup E$ and $\beta=\inf E$ are in $\overline{E}$.

Proof: If $\alpha \in E$, then $\alpha \in \overline{E}$. Suppose $\alpha \notin E$, then for each $n \in \mathbb{N}$, there exists $p_n \in E$ such that $\alpha - 1/n < p_n < \alpha$. Since, $p_n \to \alpha$, $\alpha \in E'$. Thus, $\alpha \in \overline{E}$. Similarly, $\beta \in \overline{E}$.

3.1E5 Example: Let E be a subset of $\mathbb{R}$. Show that the set of limit points of E is closed.

Suppose $p \in (E')^c$. Then there exists $\varepsilon > 0$ such that

$$N_\varepsilon(p) \cap E = \phi.$$

If $q \in N_\varepsilon(p)$, then there exists $\delta > 0$ such that

$$N_\delta(q) \subseteq N_\varepsilon(p) \Rightarrow N_\delta(q) \cap E = \phi.$$

Thus, $N_\varepsilon(p) \subseteq (E')^c$. Hence, $(E')^c$ is open and E' is closed.

3.1D4 Definition: Dense Sets

Let E be a subset of $\mathbb{R}$. E is said to be dense in $\mathbb{R}$ if $\overline{E} = \mathbb{R}$.

Note: $\mathbb{Q}$ and $\mathbb{R} \setminus \mathbb{Q}$ are dense in $\mathbb{R}$.

3.1A1 Assignment: Let E be a subset of $\mathbb{R}$. Prove that

(a) E is open if and only if $E^\circ = E$.

(b) If $O \subseteq E$ and O is open, then $O \subseteq E^\circ$.

(c) If E is dense in $\mathbb{R}$, then for every $p \in \mathbb{R}$ there exists a sequence $\{p_n\}$ in E such that $\lim_{n \to \infty} p_n = p$.

3.1D4 Definition:

Let X be a subset of $\mathbb{R}$ and let O and C be subsets of X.

(a) O is said to be open in X (or open relative to X) if for every $p \in O$, there exists $\varepsilon > 0$ such that $N_\varepsilon(p) \cap X \subseteq O$.

(b) C is said to be closed in X (or closed relative to X) if $X \setminus C$ is open in X.

3.1E6 Example: It is easily seen that $(-1, 0]$ is open in $(-\infty, 0]$ but is not open in $\mathbb{R}$.

3.1T8 Theorem:

Let X be a subset of $\mathbb{R}$ and let O and C be subsets of X.

(a) O is open in X if and only if $O = X \cap E$ for some open set E of $\mathbb{R}$.

(b) C is closed in X if and only if $C = X \cap F$ for some closed set F of $\mathbb{R}$.

Proof: Exercise.

3.1D5 Definition: Connected Sets

Let E be a subset of $\mathbb{R}$.

(a) E is said to be **disconnected** if there exist two nonempty disjoint open sets U and V such that

$$E \cap U \neq \phi \neq E \cap V$$

and

$$E = (E \cap U) \cup (E \cap V) = E \cap (U \cup V).$$

(b) E is said to be **connected** if it is not disconnected.

Note: The set $E = (0, 1 - 1/k] \cup (1, 2)$, where $k \in (\mathbb{N} \setminus \{1\})$ is fixed, is clearly disconnected because $U = (0,1)$ and $V = (1,2)$ implies that U, V are open, $U \cap V = \phi$,

$$V \cap E = (1,2) \neq \phi,\ E \cap (U \cup V) = (E \cap U) \cup (E \cap V) = (0, 1 - 1/k] \cup (1,2) = E.$$

Note: If E consists of some points greater than a and some points less than a, then we can take $U = (-\infty, a)$, $V = (a, \infty)$. In this case,

$$U \cap V = \phi,\ E \cap U \neq \phi \neq E \cap V \text{ and } E = E \cap (U \cup V).$$

3.1T9 Theorem:

A subset E of $\mathbb{R}$ is connected if and only if it is an interval.

Proof: Suppose E is disconnected. Then there exists two nonempty disjoint open sets U and V such that

$$U_1 = E \cap U \neq \phi \neq E \cap V = V_1 \text{ and } E = U_1 \cup V_1.$$

Choose $x \in U_1$ and $y \in V_1$ and assume (without loss of generality) that $x < y$. Then

$$U_1 \cap [x, y] \neq \phi \text{ and } U_1 \cap [x, y] \text{ is bounded above.}$$

Define $z = \sup(U_1 \cap [x, y])$. Then by (3.1T7), $z \in \overline{(U_1 \cap [x, y])}$. By example (3.1E4(b)),

$$\overline{(U_1 \cap [x, y])} \subseteq \bar{U}_1 \cap [x.y] \Rightarrow z \in \bar{U}_1.$$

If $z \in V$, then there exists $\varepsilon > 0$ such that $N_\varepsilon(z) \subseteq V$. But since $U \cap V = \phi$,

$$z \notin U' \text{ and } z \notin U \Rightarrow z \notin \bar{U} \Rightarrow z \notin \bar{U}_1.$$

This is a contradiction. Thus, $z \notin V$ and $z \notin V_1$. Therefore, $x \leq z < y$. If $z \notin U$, then

$$z \notin U_1 \Rightarrow x < z < y \text{ and } z \notin E.$$

If $z \in U$, then $z \notin \bar{V}$. Hence, there exists w such that

$$x \leq z < w < y$$

and

$$w \notin V \Rightarrow w \notin V_1.$$

Furthermore,

$$z < w \Rightarrow w \notin U \Rightarrow w \notin U_1.$$

Thus,

$$x \le z < w < y \text{ with } w \notin E.$$

Therefore, E is not an interval. Hence, if E is an interval, then it must be connected.

Conversely, suppose E is not an interval. Then there exist $x, y \in E$ and $z \in (x, y)$ such that $z \notin E$. Let $U = (-\infty, z)$ and V$=(z, \infty)$. Then U and V are nonempty disjoint open sets such that $E \cap U \ne \phi \ne E \cap V$ and $E = E \cap (U \cup V)$. Thus E is disconnected. Hence result.

3.1D6 Definition: Open Cover

Let E be a subset of $\mathbb{R}$. A collection $\{O_\alpha\}_{\alpha \in A}$ of open sets in $\mathbb{R}$ is said to be an **open cover** of E if $E \subseteq \bigcup_{\alpha \in A} O_\alpha$.

3.1E7 Example: Let $E = (0,1)$ and $O_n = \left(\frac{1}{n}, 1 - \frac{1}{n}\right)$, $n \in \mathbb{N}$. Suppose $p \in E = (0,1)$. If $\varepsilon = \min\{p, 1-p\}$, then there exists $n \in \mathbb{N}$ such that $0 < \frac{1}{n} < \varepsilon$. Therefore,

$$\frac{1}{n} < \varepsilon \le p \text{ and } \varepsilon \le 1 - p \Rightarrow p \le 1 - \varepsilon < 1 - \frac{1}{n}.$$

Thus

$$p \in O_n \Rightarrow p \in \bigcup_{n=1}^{\infty} O_n.$$

Hence, $\{O_n\}_{n=1}^{\infty}$ is an open cover of E.

3.1D6 Definition: Compact Sets

A subset K of $\mathbb{R}$ is said to be **compact** if every open cover of K has a finite subcover of K; i.e., if $\{O_\alpha\}_{\alpha \in A}$ is an open cover of K, then there exists $\alpha_1, \dots, \alpha_n \in A$ such that $K \subseteq \bigcup_{k=1}^{n} O_{\alpha_k}$.

Note: We can use the definition to show that a set is not compact in most cases. However, it is not easy to show that a given set is compact. We need to find sufficient conditions for this.

3.1E8 Example: Show that the open interval $E = (0,1)$ is not compact.

We have shown in example (3.1E7), that the collection $\{O_n\}_{n=1}^{\infty}$ is an open cover of E, where $O_n = \left(\frac{1}{n}, 1 - \frac{1}{n}\right)$, $n \in \mathbb{N}$. Suppose E is compact. Then there exists a finite subcover

$\{O_{n_1}, O_{n_2}, ..., O_{n_k}\}$ of E. Let $m = \max\{n_1, n_2, ..., n_k\}$. Then $O_{n_j} \subseteq O_m$ for $j = 1, 2, ..., k$ and $\bigcup_{j=1}^{k} O_{n_j} = O_m$. But

$$E = (0,1) \subseteq \bigcup_{j=1}^{k} O_{n_j} = \left(\frac{1}{m}, 1 - \frac{1}{m}\right)$$

is a contradiction. Hence, E is not compact.

3.1.T10 Theorem:

Let K be a compact subset of $\mathbb{R}$ and let F be a closed subset of K. Then

(a) K is closed and bounded

(b) F is compact.

Proof: (a) $\{O_n\}_{n=1}^{\infty}$ defined by $O_n = (-n, n)$, $n \in \mathbb{N}$ is an open cover of $\mathbb{R}$ and must also be an open cover of K. Since K is compact, there exists a finite subcover $\{O_{n_1}, O_{n_2}, ..., O_{n_k}\}$ of K. Let $m = \max\{n_1, n_2, ..., n_k\}$. Then $O_{n_j} \subseteq O_m$ for $j = 1, 2, ..., k$ and $\bigcup_{j=1}^{k} O_{n_j} = O_m$. Thus, $K \subseteq (-m, m)$ and K is bounded. Let $p \in K^c$ and for each $q \in K$, let $\varepsilon_q = |p - q|/2$. Then $N_{\varepsilon_q}(q) \cap N_{\varepsilon_q}(p) = \phi$ and $\{N_{\varepsilon_q}(q)\}_{q \in K}$ is an open cover of K. Since K is compact there exists a finite subcover $\{N_{\varepsilon_{q1}}, N_{\varepsilon_{q2}}, ..., N_{\varepsilon_{qk}}\}$ of K. Let $\varepsilon = \min\{\varepsilon_{q_1}, \varepsilon_{q_2}, ..., \varepsilon_{q_k}\} > 0$. Then

$$N_{\varepsilon_{qj}}(q) \cap N_{\varepsilon}(p) = \phi \text{ for } j = 1, 2, ..., k \text{ and } K \cap N_{\varepsilon}(p) = \phi$$

$$\Rightarrow N_{\varepsilon}(p) \subseteq K^c.$$

Hence, K^c is open and K is closed.

(b) Let $\{O_\alpha\}_{\alpha \in A}$ be an open cover of F. Then $\{O_\alpha\}_{\alpha \in A} \cup \{F^c\}$ is an open cover of K. Since K is compact, there exists a finite subcover $\{O_{n_1}, O_{n_2}, ..., O_{n_k}, F^c\}$ of K. Thus, $\{O_{n_1}, O_{n_2}, ..., O_{n_k}\}$ is a finite subcover of F. Hence, F is compact.

Note: From the above theorem, we note that any set that is not closed cannot be compact; also any unbounded set cannot be compact.

3.1E9 Example:

(a) $\left\{\frac{1}{n} \middle| n \in \mathbb{N}\right\}$ is not compact because it is not closed, but $\left\{\frac{1}{n} \middle| n \in \mathbb{N}\right\} \cup \{0\}$ is compact.

(b) $E=\{x\in\mathbb{R}|x\geq 0\}$ is not compact because it is not bounded, even though it is closed.

3.1T11 Theorem:

Suppose S is an infinite subset of a compact set K. Then S has a limit point in K.

Proof: Suppose no point of K is a limit point of S. Then for each $q\in K$, there exists a neighborhood $N_\varepsilon(q)$ of q such that $N_\varepsilon'(q)\cap S=\phi$. The collection $\{N_\varepsilon(q)|q\in K\}$ is an open cover of K and also of S. Since S is infinite, no finite subcover of the above open cover can cover S. Thus, no finite subcover of the above cover can cover K. This is a contradiction. Hence, S has a limit point in K.

3.1T12 Theorem: Heine-Borel Theorem

Every closed and bounded interval $[a,b]$ of $\mathbb{R}$ is compact.

Proof: Let $\{O_\alpha\}_{\alpha\in A}$ be an open cover of $[a,b]$ and let $E=\{r\in[a,b]|[a,r]\subseteq U_n\}$, where $U_n=\bigcup_{k=1}^{n}O_{\alpha_k}$ and $\alpha_1,\alpha_2,\ldots,\alpha_n\in A$ for some $n\in\mathbb{N}$. Then E is bounded above by b and must have a supremum, say c, in $\mathbb{R}$. Since b is an upper bound of E, $c\leq b$. Since $c\in[a,b]$, $c\in O_\beta$ for some $\beta\in A$.

$$O_\beta \text{ open} \Rightarrow \text{there exists } \varepsilon>0 \text{ such that } (c-\varepsilon,c+\varepsilon)\subseteq O_\beta.$$

Since $c-\varepsilon$ is not an upper bound of E, there exists $r\in E$ such that $c-\varepsilon<r\leq c$. Since $[a,r]\subset U_n$, the finite union $U_n\cup O_\beta$ contains $[a,c]$. Therefore, $c\in E$. If $c<b$, choose $s<b$ such that $c<s<c+\varepsilon$. Then the finite union $U_n\cup O_\beta$ also contains $[a,s]$. Thus $s\in E$, contradicting $c=\sup E$. Hence, $c=b$.

3.1T13 Theorem: Heine-Borel-Bolzano-Weierstrass

Let K be a subset of $\mathbb{R}$. Then the following statements are equivalent:

(a) K is closed and bounded.
(b) K is compact.
(c) Every infinite subset of K has a limit point in K.

Proof: If (a) holds, then there exists $M>0$ such that $-M\leq x\leq M$ for all $x\in K$. Since K is closed and contained in a compact set, it must also be compact (3.1T10 (b)). This proves that (a) $\Rightarrow$ (b).

(b) $\Rightarrow$ (c) is Theorem (3.1T11).

Suppose (c) holds. If K is not bounded, then for every $n\in\mathbb{N}$ there exists $p_n\in K$ with $p_n\neq p_m$ for $n\neq m$ such that $|p_n|>n$. Then $\{p_n|n\in\mathbb{N}\}$ is an infinite subset of $\mathbb{R}$ with no limit point in $\mathbb{R}$, and hence, no limit point in K. This is a contradiction. Thus K is bounded.

Let p be a limit point of K. Then for each $n \in \mathbb{N}$ there exists $p_n \in K$ with $p_n \neq p$ such that $|p_n - p| < 1/n$ and $p_n \neq p_m$ for $n \neq m$. Then $S = \{p_n | n \in \mathbb{N}\}$ is an infinite subset of K and p is a limit point of S. If p is the only limit point of S, then by hypothesis p is in K and K is closed by theorem (3.1T5).

Suppose $q \in \mathbb{R}$ with $q \neq p$. Let n_0 be a positive integer greater than $\frac{1}{\varepsilon} = \frac{2}{|p-q|}$. Then

$$\begin{aligned}|p_n - q| &\geq |p-q| - |p_n - q| \\ &> |p-q| - 1/n \\ &> |p-q| - \frac{|p-q|}{2} = \frac{|p-q|}{2} = \varepsilon\end{aligned}$$

for all $n \geq n_0$. Thus $N_\varepsilon(q)$ contains at most finitely many p_n's, and q cannot be a limit point of S. Thus p is the only limit point of S and K is closed. Hence, (c) $\Rightarrow$ (a). This completes the proof.

3.1T14 Theorem:

Let K be a nonempty compact subset of $\mathbb{R}$. Then every sequence $\{p_n\}$ in K has a subsequence that converges to a point in K.

Proof: Since K is compact and $\{p_n\}$ is in K, $\{p_n\}$ is a bounded sequence in $\mathbb{R}$ and must have a convergent subsequence $\{p_{n_k}\}$ (2.4T3C), with limit p. Since K is closed, p must be in K. Let $S = \{p_{n_k} | k \in \mathbb{N}\}$. If S is finite, then $p = p_{n_k}$ for some $k \in \mathbb{N}$. Thus, $p \in S \subseteq K$. If S is infinite, then p is a limit point of an infinite subset S of K, and must be in K (2.1T3).

3.1E10 Example: Let $E = \left\{\frac{1}{n} \middle| n \in \mathbb{N}\right\}$. Prove that (using (3.1D6)) (i) E is not compact. (ii) $K = E \cup \{0\}$ is compact.

(i) For each $n \in \mathbb{N}$, let $O_n = \left(\frac{1}{n} - \varepsilon_n, \frac{1}{n} + \varepsilon_n\right)$, where $\varepsilon_n = \frac{1}{n} - \frac{1}{n+1} > 0$. Then

$$O_n \cap E = \left\{\frac{1}{n}\right\} \text{ for all } n \in \mathbb{N} \text{ and } E \subseteq \bigcup_{n=1}^{\infty} O_n.$$

Suppose E is compact. Then there exists $p \in \mathbb{N}$ such that

$$E \subseteq \bigcup_{k=1}^{p} O_{n_k}.$$

Let $m = \max\{n_k | k = 1, 2, ..., p\}$. Then

$$\frac{1}{m+1} \notin \bigcup_{n=1}^{m} O_n \supseteq E.$$

This is a contradiction. Hence, E is not compact.

(ii) Let $\{O_\alpha\}_{\alpha\in A}$ be an open cover of $K = E\cup\{0\}$. Then there exists $\alpha_0 \in A$ such that $0\in O_{\alpha_0}$. Since O_{α_0} is open, there exists $\varepsilon > 0$ such that $(-\varepsilon,\varepsilon)\subseteq O_{\alpha_0}$. Now choose $n\in\mathbb{N}$ satisfying $0<\frac{1}{n}<\varepsilon$. Since $\{O_\alpha\}_{\alpha\in A}$ is an open cover of K, there exist indices $\alpha_k \in A$; $k=1,2,...,n$ such that $\frac{1}{k}\in O_{\alpha_k}$; $k=1,2,...,n$. The finite collection $\{O_{\alpha_k}\}_{k=0}^{n}$ is clearly a finite subcover of K.

3.1A2 Assignment:
Prove that a compact set can have only a finite number of isolated points.

3.2 Limit of a Function

We give a formal definition of the limit of a function in this section, and obtain an important result connecting the limit of a function to the limit of a real sequence.

3.2.D1 Definition: Limit of a Function

Let $E\subseteq\mathbb{R}$ and $f: E\to\mathbb{R}$. Suppose that p is a limit point of E and let $L\in\mathbb{R}$. The function f is said to have the **limit** L at p if for every $\varepsilon>0$ there exists $\delta>0$ such that

$$f(x)\in N_\varepsilon(L) \text{ whenever } x\in E \text{ and } x\in N_\delta'(p).$$

In this case, we write

$$\lim_{x\to p} f(x)=L \text{ or } f(x)\to L \text{ as } x\to p.$$

The above statement can also be written as: For every $\varepsilon>0$ there exists $\delta>0$ such that

$$|f(x)-L|<\varepsilon \text{ whenever } x\in E \text{ and } 0<|x-p|<\delta.$$

3.2E1 Example: Use the definition of limit to prove that:

(i) $\lim_{x\to -1}(x^2-2x-1)=2$ (ii) $\lim_{x\to 2}\sqrt{x^2+5}=3$ (iii) $\lim_{x\to -2}\frac{2x-1}{x^2+1}=-1.$

(i)
$$\begin{aligned}|(x^2-2x-1)-2| &= |x^2-2x-3|\\ &= |x-3||x+1| < 3|x+1|\end{aligned}$$

whenever $|x-(-1)|<1$. Note that

$$\begin{aligned}|x-(-1)|=|x+1|<1 &\Leftrightarrow 0<x<2\\ &\Leftrightarrow -3<x-3<-1\\ &\Rightarrow |x-3|<3.\end{aligned}$$

Let $\varepsilon > 0$ be arbitrary and $\delta = \min\{1, \varepsilon/3\}$. Then

$$\begin{aligned}\left|\left(x^2 - 2x - 1\right) - 2\right| &< 3|x+1| \\ &< 3\cdot(\varepsilon/3) = \varepsilon\end{aligned}$$

whenever $0 < |x - (-1)| < \delta$.

(ii) $$\left|\sqrt{x^2+5} - 3\right| = \left|\frac{\left(\sqrt{x^2+5}-3\right)\left(\sqrt{x^2+5}+3\right)}{\sqrt{x^2+5}+3}\right| = \left|\frac{x^2-4}{\sqrt{x^2+5}+3}\right| = \frac{|x-2||x+2|}{\sqrt{x^2+5}+3}.$$

$$|x-2| < 1 \Leftrightarrow 1 < x < 3 \Leftrightarrow 3 < x+2 < 5 \Rightarrow |x+2| < 5.$$

$$\begin{aligned}|x-2| < 1 \Leftrightarrow 1 < x < 3 &\Leftrightarrow \sqrt{6}+3 < \sqrt{x^2+5}+3 < \sqrt{14}+3 \\ &\Rightarrow \sqrt{x^2+5}+3 > \sqrt{6}+3 > 5.\end{aligned}$$

Therefore,

$$\begin{aligned}\left|\sqrt{x^2+5} - 3\right| &= \frac{|x-2||x+2|}{\sqrt{x^2+5}+3} \\ &< \frac{|x-2|5}{5} = |x-2|\end{aligned}$$

if $|x-2| < 1$. Let $\varepsilon > 0$ be arbitrary and $\delta = \min\{1, \varepsilon\}$. Then

$$\left|\sqrt{x^2+5} - 3\right| < |x-2| < \varepsilon$$

whenever $0 < |x-2| < \delta$.

(iii) $$\begin{aligned}\left|\frac{2x-1}{x^2+1} - (-1)\right| &= \left|\frac{x^2+2x}{x^2+1}\right| \\ &= \frac{|x||x-(-2)|}{x^2+1} < \frac{3}{2}|x-(-2)|\end{aligned}$$

whenever $|x-(-2)| < 1$. Note that

$$\begin{aligned}|x-(-2)| = |x+2| < 1 &\Leftrightarrow -3 < x < -1 \\ &\Rightarrow 1 < |x| < 3 \text{ and } x^2+1 > 2.\end{aligned}$$

Let $\varepsilon > 0$ be arbitrary and $\delta = \min\{1, 2\varepsilon/3\}$. Then

$$\begin{aligned}\left|\frac{2x-1}{x^2+1} - (-1)\right| &< \frac{3}{2}|x-(-2)| \\ &< \frac{3}{2}\cdot\frac{2\varepsilon}{3} = \varepsilon\end{aligned}$$

whenever $0<|x-(-2)|<\delta$.

Note: In some of the above examples, finding δ in terms of ε was rather tedious.

3.2A1 Assignment: Use the definition of limit to prove that:

(i) $\lim_{x\to 2}\left(x^2-2x+1\right)=1$ (ii) $\lim_{x\to 0}\sqrt{3x^2+4}=2$ (iii) $\lim_{x\to -2}\dfrac{x-1}{x+1}=3$.

3.2E2 Example: Use the definition of limit to prove that $\lim_{x\to 1} f(x)=2$ but $\lim_{x\to p} f(x)$ does not exist for any $p\in(\mathbb{R}\setminus\{1\})$ where $f:\mathbb{R}\to\mathbb{R}$ is defined by

$$f(x)=\begin{cases}2x & \text{if } x\in\mathbb{Q}\\ 3-x & \text{if } x\notin\mathbb{Q}.\end{cases}$$

If $x\in\mathbb{Q}$, then

$$|f(x)-2|=|2x-2|=2|x-1|.$$

If $x\notin\mathbb{Q}$, then

$$\begin{aligned}|f(x)-2|&=|(3-x)-2|\\&=|1-x|=|x-1|.\end{aligned}$$

Let $\varepsilon>0$ be arbitrary and $\delta=\varepsilon/2$. Then

$$|f(x)-1|\le 2|x-1|<2\cdot\varepsilon/2=\varepsilon$$

whenever $0<|x-1|<\delta$. Suppose $\lim_{x\to p} f(x)=L$ where $p\in(\mathbb{R}\setminus\{1\})$ and $L\in\mathbb{R}$. Then $|2p-L|$ and $|3-p-L|$ cannot both be zero. If $|2p-L|\le|3-p-L|$, let $2\varepsilon=|3-p-L|>0$. Then there exists $\delta>0$ such that

$$|f(x)-L|<\varepsilon \text{ whenever } 0<|x-p|<\delta.$$

Now choose $q\in(\mathbb{R}\setminus\mathbb{Q})$ such that $0<|q-p|<\min\{\varepsilon,\delta\}$. Then

$$\begin{aligned}|f(q)-L|&=|(3-q)-L|\\&=|3-p-L+p-q|\\&\ge\big||3-p-L|-|p-q|\big|\\&>2\varepsilon-\varepsilon=\varepsilon.\end{aligned}$$

This is a contradiction.

If $|2p-L|>|3-p-L|$, let $3\varepsilon=|2p-L|>0$. Then there exists $\delta>0$ such that

$$|f(x)-L|<\varepsilon \text{ whenever } 0<|x-p|<\delta.$$

Choose $q\in\mathbb{Q}$ such that $0<|q-p|<\min\{\varepsilon,\delta\}$. Then

$$\begin{aligned}|f(q)-L| &= |2q-L| \\ &= |2p-L+2(q-p)| \\ &\geq \big||2p-L|-2|p-q|\big| \\ &> 3\varepsilon-2\varepsilon=\varepsilon.\end{aligned}$$

This is also a contradiction. Hence, $\lim_{x\to p} f(x)$ does not exist if $p\in(\mathbb{R}\setminus\{1\})$.

Note: From the above example, we see that it is not easy to prove that $\lim_{x\to p} f(x)$ does not exist. We should develop theorems on limits to make or tasks a little easier.

3.2.T1 Theorem:

Let $E\subseteq\mathbb{R}$ and $f: E\to\mathbb{R}$. Suppose that p is a limit point of E and let $L\in\mathbb{R}$. Then

$$\lim_{x\to p} f(x)=L \text{ if and only if } \lim_{n\to\infty} f(p_n)=L$$

for every sequence $\{p_n\}$ in E, with $p_n\neq p$ for all $n\in\mathbb{N}$ and $\lim_{n\to\infty} p_n=p$.

Proof: Suppose $\lim_{x\to p} f(x)=L$ and let $\varepsilon>0$ be given. Then there exists $\delta>0$ such that

$$|f(x)-L|<\varepsilon \text{ whenever } x\in E \text{ and } 0<|x-p|<\delta.$$

If $\{p_n\}$ is any sequence in E with limit p, then there exists $n_0\in\mathbb{N}$ such that

$$|p_n-p|<\delta \text{ for all } n\geq n_0.$$

Therefore, for all $n\geq n_0$, $|f(p_n)-L|<\varepsilon$. Hence, $\lim_{n\to\infty} f(p_n)=L$.

Suppose $\lim_{x\to p} f(x)\neq L$. Then there exists $\varepsilon>0$ such that for all $\delta>0$ there exists $x\in E$ with $0<|x-p|<\delta$ and

$$|f(x)-L|\geq\varepsilon.$$

Now for each $n_0\in\mathbb{N}$ let $\delta=\frac{1}{n}\cdot$ Then there exists $p_n=x\in E$ such that $0<|p_n-p|<\delta$ and

$$|f(p_n)-L|\geq\varepsilon.$$

Thus, there exists a sequence $\{p_n\}$ in E, with $p_n \neq p$ for all $n \in \mathbb{N}$ and $\lim_{n\to\infty} p_n = p$; but $\{f(p_n)\}$ does not converge to L. Hence, if $\lim_{n\to\infty} f(p_n) = L$ for every sequence $\{p_n\}$ in E, with $p_n \neq p$ for all $n \in \mathbb{N}$ and $\lim_{n\to\infty} p_n = p$, then $\lim_{x\to p} f(x) = L$.

3.2T1C Corollary:

If $\lim_{x\to p} f(x)$ exists, then it is unique.

Proof: Follows immediately because, $\lim_{n\to\infty} f(p_n) = L$ and $\lim_{n\to\infty} f(p_n) = L' \Rightarrow L = L'$.

3.2E3 Example: Use theorem (3.2T1), to prove that $\lim_{x\to p} f(x)$ does not exist for any $p \in (\mathbb{R} \setminus \{1\})$ where $f : \mathbb{R} \to \mathbb{R}$ is defined as $f(x) = \begin{cases} 2x & \text{if } x \in \mathbb{Q} \\ 3-x & \text{if } x \notin \mathbb{Q}. \end{cases}$

This is the same as example (3.2E2). But this proof is a lot easier. Suppose $p \in (\mathbb{R} \setminus \{1\})$. Let $\{p_n\}$ be a sequence in $\mathbb{Q}$ and $\{q_n\}$ be a sequence in $\mathbb{R} \setminus \mathbb{Q}$ such that

$$\lim_{n\to\infty} p_n = p = \lim_{n\to\infty} q_n.$$

Then

$$f(p_n) = 2p_n \to 2p \text{ and } f(q_n) = 3 - p_n \to 3 - p.$$

But $p \neq 1$ implies that $2p \neq 3 - p$. Hence,

$$\lim_{n\to\infty} f(p_n) \neq \lim_{n\to\infty} f(q_n)$$

and $\lim_{x\to p} f(x)$ does not exist for any $p \neq 1$.

3.2T2 Theorem: Limit Theorems

Suppose $E \subseteq \mathbb{R}$ and $f, g : E \to \mathbb{R}$, p is a limit point of E, $\lim_{x\to p} f(x) = A$, and $\lim_{x\to p} g(x) = B$, where $A, B \in \mathbb{R}$. Then

(a) $\lim_{x\to p}[f(x) + g(x)] = A + B$

(b) $\lim_{x\to p}[f(x)g(x)] = AB$

(c) $\lim_{x\to p} \dfrac{f(x)}{g(x)} = \dfrac{A}{B}$ provided $B \neq 0$.

Proof: These results follow from theorem (3.2T1) and analogous results for sequences.

3.2D2 Definition: Bounded Functions

Let $E \subseteq \mathbb{R}$ and $f : E \to \mathbb{R}$. Then f is said to be **bounded above** if there exists $M \in \mathbb{R}$ such that $f(x) \le M$ for all $x \in E$.

f is said to be **bounded below** if there exists $m \in \mathbb{R}$ such that $f(x) \ge m$ for all $x \in E$.

f is said to be **bounded** if there exists $M \in \mathbb{R}$, $M > 0$ such that $|f(x)| \le M$ for all $x \in E$.

Note: A real valued function is bounded if and only if it is bounded above and bounded below.

3.2T3 Theorem:

Suppose $E \subseteq \mathbb{R}$ and $f, g : E \to \mathbb{R}$, p is a limit point of E, $\lim_{x \to p} f(x) = 0$, and g is bounded. Then $\lim_{x \to p} f(x) g(x) = 0$.

Proof: Since g is bounded, there exists $M > 0$ such that $|g(x)| \le M$ for all $x \in E$. Let $\varepsilon > 0$ be given. Since $\lim_{x \to p} f(x) = 0$, there exists $\delta > 0$ such that

$$|f(x) - 0| < \varepsilon / M$$

whenever $x \in E$ and $0 < |x - p| < \delta$. This implies that

$$\begin{aligned} |f(x)g(x) - 0| &= |f(x)||g(x)| \\ &< (\varepsilon / M) M = \varepsilon \end{aligned}$$

whenever $x \in E$ and $0 < |x - p| < \delta$. Hence,

$$\lim_{x \to p} f(x) g(x) = 0.$$

3.2T4 Theorem: Squeeze Theorem

Suppose $E \subseteq \mathbb{R}$ and $f, g, h : E \to \mathbb{R}$, p is a limit point of E, and

$$f(x) \le g(x) \le h(x) \text{ for all } x \in E.$$

If $\lim_{x \to p} f(x) = \lim_{x \to p} h(x) = L$, then $\lim_{x \to p} g(x) = L$.

Proof: This result follows from theorem (3.2T1) and an analogous result for sequences.

3.2T5 Theorem:

Suppose $E \subseteq \mathbb{R}$ and $f : E \to \mathbb{R}$, and p is a limit point of E. Then $\lim_{x \to p} f(x) = 0$ if and only if $\lim_{x \to p} |f(x)| = 0$.

Proof: This result follows from theorem (3.2T1) and an analogous result for sequences.

3.2T6 Theorem:

Suppose $E \subseteq \mathbb{R}$ and $f : E \to \mathbb{R}$, and p is a limit point of E. If $\lim_{x \to p} f(x) = L > 0$, then $f(x) > L/2$ for all $x \in N_\delta'(p)$ for some $\delta > 0$.

Proof: This result follows from theorem (3.2T1) and an analogous result for sequences.

Note: In the previous theorem, we may replace the phrase

$$\lim_{x \to p} f(x) = L > 0, \text{ then } f(x) > L/2$$

with the phrase

$$\lim_{x \to p} f(x) = L < 0, \text{ then } f(x) < L/2.$$

3.2E4 Example: Let f be a polynomial of degree n, i.e., $f(x) = \sum_{k=0}^{n} a_k x^k$, where $a_k, x \in \mathbb{R}$ for $k = 0, 1, \ldots, n$ with $n \in \mathbb{N} \cup \{0\}$ and $p, x \in \mathbb{R}$. Show that $\lim_{x \to p} f(x) = \sum_{k=0}^{n} a_k p^k = f(p)$.

It is easily seen that $\lim_{x \to p} a = \lim_{x \to p} ax^0 = a$ for any $a \in \mathbb{R}$ and $\lim_{x \to p} x^1 = \lim_{x \to p} x = p$. Using the principle of mathematical induction, we can show that $\lim_{x \to p} x^k = p^k$ for all $k \in \mathbb{N}$, since

$$\lim_{x \to p} x^{k+1} = \lim_{x \to p} x^k \lim_{x \to p} x^1 = p^k p = p^{k+1} \text{ if } \lim_{x \to p} x^k = p^k.$$

Let $P(n)$: $\lim_{x \to p}\left(\sum_{k=0}^{n} a_k x^k\right) = \sum_{k=0}^{n} a_k p^k$ for $n = 0, 1, 2, \ldots$ Since, $\lim_{x \to p} a_0 = a_0$, $P(0)$ is true.

Suppose $P(n)$ is true for some $n \in \mathbb{N} \cup \{0\}$. Then

$$\lim_{x \to p}\left(\sum_{k=0}^{n+1} a_k x^k\right) = \lim_{x \to p}\left(x \cdot a_{n+1} x^n + \sum_{k=0}^{n} a_k x^k\right) = p \cdot a_{n+1} p^n + \sum_{k=0}^{n} a_k p^k = \sum_{k=0}^{n+1} a_k p^k,$$

i.e., $P(n+1)$ is true. Therefore, $P(n)$ is true for all $n \in \mathbb{N} \cup \{0\}$. Hence,

$$\lim_{x \to p} f(x) = \sum_{k=0}^{n} a_k p^k = f(p).$$

3.2E5 Example: Let f be a rational function, i.e., $f(x) = \frac{p(x)}{q(x)}$, where p and q are polynomials and $x \in \mathbb{R}$ such that $q(x) \neq 0$. Show that $\lim_{x \to c} f(x) = f(c)$ for all $c \in \mathbb{R}$ with $q(c) \neq 0$.

Let $c \in \mathbb{R}$ with $q(c) \neq 0$. Then $q(x) \neq 0$ for all $x \in N_\delta'(c)$ for some $\delta > 0$ (3.2T6). Thus, $N_\delta'(c) \subseteq Dom(f)$. Therefore, c is a limit point of $Dom(f)$ and by theorem (3.2T2) and (3.2E4) above,

$$\lim_{x \to c} f(x) = \frac{\lim_{x \to c} p(x)}{\lim_{x \to c} q(x)} = \frac{p(c)}{q(c)} = f(c).$$

Note: The limit of a polynomial or a rational function is the function value at that point.

3.2A2 Assignment: Suppose $f: E \to \mathbb{R}$, and p is a limit point of E, and $\lim_{x \to p} f(x) = L$.

(a) Prove that $\lim_{x \to p} |f(x)| = |L|$.

(b) If $f(x) \geq 0$ for all $x \in E$, then prove that $\lim_{x \to p} \sqrt{f(x)} = \sqrt{L}$.

(c) Prove that $\lim_{x \to p} (f(x))^n = L^n$ for all $n \in \mathbb{N}$.

(d) Prove that if $L \neq 0$ and $\lim_{x \to p} f(x)g(x)$ exists, then $\lim_{x \to p} g(x)$ exists.

3.2D3 Definition: Limits at Infinity

(a) Suppose $f: E \to \mathbb{R}$, where E is not bounded above and $L \in \mathbb{R}$. Then f is said to have a limit L at ∞ if for every $\varepsilon > 0$ there exists $M > 0$ such that

$$|f(x) - L| < \varepsilon \text{ for all } x \in E \cap (M, \infty).$$

In this case, we write: $\lim_{x \to \infty} f(x) = L$ or $f(x) \to L$ as $x \to \infty$.

(b) Suppose $f: E \to \mathbb{R}$, where E is not bounded above and $L \in \mathbb{R}$. Then f is said to have a limit L at $-\infty$ if for every $\varepsilon > 0$ there exists $M > 0$ such that

$$|f(x) - L| < \varepsilon \text{ for all } x \in E \cap (-\infty, -M).$$

In this case, we write: $\lim_{x \to -\infty} f(x) = L$ or $f(x) \to L$ as $x \to -\infty$.

Note: Theorems (3.2T1 – 3.2T6) remain valid if we replace p by $-\infty$ or ∞, with appropriate changes in the hypotheses of these theorems.

3.2D4 Definition: Infinite Limits

Suppose $f: E \to \mathbb{R}$ and p is a limit point of E.

(a) The function f is said to have limit ∞ at p if for every $X > 0$ there exists $\delta > 0$ such that

$$f(x) > X \text{ whenever } x \in E \text{ and } x \in N_\delta'(p).$$

In this case, we write $\lim_{x \to p} f(x) = \infty$ or $f(x) \to \infty$ as $x \to p$.

(b) The function f is said to have limit $-\infty$ at p if for every $X > 0$ there exists $\delta > 0$ such that

$$f(x) < -X \text{ whenever } x \in E \text{ and } x \in N_\delta'(p).$$

In this case, we write $\lim_{x \to p} f(x) = -\infty$ or $f(x) \to -\infty$ as $x \to p$.

3.2D5 Definition: Infinite Limits at Infinity

(a) Suppose $f : E \to \mathbb{R}$ and E is not bounded above. Then f is said to have a limit ∞ at ∞ if for every $X > 0$ there exists $M > 0$ such that

$$f(x) > X \text{ for all } x \in E \cap (M, \infty).$$

In this case, we write $\lim_{x\to\infty} f(x) = \infty$ or $f(x) \to \infty$ as $x \to \infty$.

(b) Suppose $f : E \to \mathbb{R}$ and E is not bounded below. Then f is said to have a limit ∞ at $-\infty$ if for every $X > 0$ there exists $M > 0$ such that

$$f(x) > X \text{ for all } x \in E \cap (-\infty, M).$$

In this case, we write $\lim_{x\to-\infty} f(x) = \infty$ or $f(x) \to \infty$ as $x \to -\infty$.

(c) Suppose $f : E \to \mathbb{R}$ and E is not bounded above. Then f is said to have a limit $-\infty$ at ∞ if for every $X > 0$ there exists $M > 0$ such that

$$f(x) < -X \text{ for all } x \in E \cap (M, \infty).$$

In this case, we write $\lim_{x\to\infty} f(x) = -\infty$ or $f(x) \to -\infty$ as $x \to \infty$.

(d) Suppose $f : E \to \mathbb{R}$ and E is not bounded below. Then f is said to have a limit $-\infty$ at $-\infty$ if for every $X > 0$ there exists $M > 0$ such that

$$f(x) < -X \text{ for all } x \in E \cap (-\infty, M).$$

In this case, we write $\lim_{x\to-\infty} f(x) = -\infty$ or $f(x) \to -\infty$ as $x \to -\infty$.

Note: For the limits in definitions (3.5D3 – 3.5D5), we can present modified versions of theorem (3.2T1), e.g.,

(a) Let $E \subseteq \mathbb{R}$ and $f : E \to \mathbb{R}$. Suppose that p is a limit point of E. Then $\lim_{x\to p} f(x) = \infty$ if and only if $\lim_{n\to\infty} f(p_n) = \infty$ for every sequence $\{p_n\}$ in E, with $p_n \neq p$ for all $n \in \mathbb{N}$ and $\lim_{n\to\infty} p_n = p$.

(b) Let $E \subseteq \mathbb{R}$ and $f : E \to \mathbb{R}$. Then $\lim_{x\to\infty} f(x) = L$ if and only if $\lim_{n\to\infty} f(p_n) = \infty$ for every sequence $\{p_n\}$ in E with $\lim_{n\to\infty} p_n = \infty$.

(c) Let $E \subseteq \mathbb{R}$ and $f : E \to \mathbb{R}$. Then $\lim_{x\to-\infty} f(x) = -\infty$ if and only if $\lim_{n\to\infty} f(p_n) = -\infty$ for every sequence $\{p_n\}$ in E with $\lim_{n\to\infty} p_n = -\infty$.

3.2T7 Theorem:

Suppose $f, g : E \to \mathbb{R}$, $\lim_{x\to\infty} f(x) = L$, and $\lim_{x\to\infty} g(x) = \infty$. Then

(a) $\lim_{x\to\infty}\left[g(x)+f(x)\right] = \infty$.

(b) $\lim_{x\to\infty}\left[g(x)f(x)\right] = \infty$ if $L > 0$.

(c) $\lim_{x\to\infty}\left[g(x)f(x)\right] = -\infty$ if $L < 0$.

(d) $\lim_{x\to\infty}\dfrac{f(x)}{g(x)} = 0$.

Proof:

(a) Let $X > 0$ be given. Then there exists $M_1, M_2 > 0$ such that

$$|f(x)-L| < 1 \text{ for all } x \in E \cap (M_1, \infty)$$

and

$$g(x) > X + |L| + 1 \text{ for all } x \in E \cap (M_2, \infty).$$

Let $M = \max\{M_1, M_2\} > 0$. Then

$$g(x)+f(x) > X + |L| + 1 + (-1-|L|) = X$$

for all $x \in E \cap (M, \infty)$. Thus $\lim_{x\to\infty}\left[g(x)+f(x)\right] = \infty$.

(b) Let $X > 0$ be given. Then there exists $M_1, M_2 > 0$ such that

$$|f(x)-L| < L/2 \text{ for all } x \in E \cap (M_1, \infty)$$

and

$$g(x) > 2X/L \text{ for all } x \in E \cap (M_2, \infty).$$

Let $M = \max\{M_1, M_2\} > 0$. Then

$$g(x)f(x) > (2X/L)(L/2) = X$$

for all $x \in E \cap (M, \infty)$. Thus $\lim_{x\to\infty}\left[g(x)f(x)\right] = \infty$.

(c) Exercise.

(d) Let $\varepsilon > 0$ be given. Then there exists $M_1, M_2 > 0$ such that

$$|f(x)-L| < 1 + |L| \text{ for all } x \in E \cap (M_1, \infty)$$

and

$$g(x) > (1+2|L|)/\varepsilon \text{ for all } x \in E \cap (M_2, \infty).$$

Let $M = \max\{M_1, M_2\} > 0$. Then

$$\left|\frac{f(x)}{g(x)}-0\right|=\frac{|f(x)|}{|g(x)|}<(1+2|L|)\cdot\frac{\varepsilon}{1+2|L|}=\varepsilon$$

for all $x \in E \cap (M,\infty)$. Thus $\lim_{x\to\infty}\frac{f(x)}{g(x)}=0.$

Note: The above theorem remains valid if we replace $x \to \infty$ by $x \to -\infty$.

3.2E6 Example: Investigate the limit at ∞ of each of the following functions:

(i) $f(x)=x^p,\ p\in\mathbb{R}$ (ii) $f(x)=\frac{p(x)}{q(x)}$, where p and q are polynomials.

(i) If $p=0$, then $f(x)=1$ for all $x\in\mathbb{R}$ and $\lim_{x\to\infty} f(x)=1.$

Suppose $p>0$ and let $X>0$ be given. If $M=X^{1/p}$, then $x>M$ implies that

$$f(x)=x^p>M^p=\left(X^{1/p}\right)^p=X.$$

Thus, $\lim_{x\to\infty} f(x)=\infty.$

Suppose $p<0$ and let $\varepsilon>0$ be given. If $M=\varepsilon^{1/p}$, then $x>M$ implies that

$$|f(x)-0|=x^p<\left(\varepsilon^{1/p}\right)^p=\varepsilon.$$

(Note: if $p<0$ and $0<y<z$, then $y^p>z^p$.) Thus, $\lim_{x\to\infty} f(x)=0.$

(ii) Let $p(x)=\sum_{k=0}^{m} a_k x^k$ and $q(x)=\sum_{k=0}^{n} b_k x^k$, where $a_k\in\mathbb{R}$ for $k=0,1,\ldots,m,$ $b_k\in\mathbb{R}$ for $k=0,1,\ldots,n,$ with $m,n\in\mathbb{N}\cup\{0\}$ and $x\in\mathbb{R}$. There are three cases to be considered.

Suppose $m<n$. Then

$$f(x)=\frac{\sum_{k=0}^{m} a_k x^k}{\sum_{k=0}^{n} b_k x^k}=\frac{\sum_{k=0}^{m} a_k x^{k-n}}{\sum_{k=0}^{n} b_k x^{k-n}}\to\frac{\sum_{k=0}^{m} 0}{b_n+\sum_{k=0}^{n-1} 0}=0 \text{ as } x\to\infty,$$

i.e., $\lim_{x\to\infty} f(x)=0.$

Suppose $m=n$. Then

$$f(x)=\frac{\sum_{k=0}^{m} a_k x^k}{\sum_{k=0}^{n} b_k x^k}=\frac{\sum_{k=0}^{m} a_k x^{k-n}}{\sum_{k=0}^{n} b_k x^{k-n}}\to\frac{a_n+\sum_{k=0}^{m} 0}{b_n+\sum_{k=0}^{n-1} 0}=\frac{a_n}{b_n} \text{ as } x\to\infty,$$

i.e., $\lim_{x\to\infty} f(x)=\frac{a_n}{b_n}.$

Suppose $m>n$. Then

$$f(x)=\frac{\sum_{k=0}^{m}a_kx^k}{\sum_{k=0}^{n}b_kx^k}=\frac{\sum_{k=0}^{m}a_kx^{k-n}}{\sum_{k=0}^{n}b_kx^{k-n}}=\frac{\frac{a_m}{b_n}\left(x^{m-n}\right)\sum_{j=0}^{m-n}\frac{a_{j+n}}{a_m}x^{j+n-m}+\sum_{k=0}^{n-1}\frac{a_k}{b_n}x^{k-n}}{1+\sum_{k=0}^{n-1}\frac{b_k}{b_n}x^{k-n}}$$

$$\to\frac{\frac{a_m}{b_n}(\infty)(1)+0}{1+\sum_{k=0}^{n-1}0}=\operatorname{sgn}\left(\frac{a_m}{b_n}\right)\infty \textit{ as } x\to\infty,$$

i.e., $\lim_{x\to\infty}f(x)=\operatorname{sgn}\left(\frac{a_m}{b_n}\right)\infty$.

Note: It is easily seen that

(i) $\lim_{x\to-\infty}f(x)=0$ if $m<n$. (ii) $\lim_{x\to-\infty}f(x)=\frac{a_n}{b_n}$ if $m=n$.

(iii) $\lim_{x\to-\infty}f(x)=(-1)^{m-n}\operatorname{sgn}\left(\frac{a_m}{b_n}\right)\infty$ if $m>n$.

3.2A3 Assignment:

(a) Determine whether the given limit exists or not. Justify your answers! It is not necessary to use the definition of limit.

(i) $\lim_{x\to0}\frac{x}{|x|}$ (ii) $\lim_{x\to0}\frac{\sqrt{x^2+1}}{|x|}$ (iii) $\lim_{x\to4}\frac{x-4}{\sqrt{x}-2}$.

(b) Determine the following limits. Justify your answers. It is not necessary to use the definition of limit.

(i) $\lim_{x\to3}\sqrt{\frac{2x-1}{x+2}}$ (ii) $\lim_{x\to0}\frac{|x+1|-|x+1|}{x}$ (iii) $\lim_{x\to\infty}\frac{x-2}{\sqrt{x^2+4}}$.

(c) Prove theorem (3.2T7(c)).

(d) Suppose $f:E\to\mathbb{R}$ and p is a limit point of E. If $\lim_{x\to p}f(x)=0$ and $f(x)>0$ for all $x\in E$, then prove that $\lim_{x\to p}\frac{1}{f(x)}=\infty$. Show by example that this conclusion cannot be made if the condition $f(x)>0$ is not there.

(e) Let $f:(0,\infty)\to\mathbb{R}$ and $\lim_{x\to\infty}\left[xf(x)\right]=L\in\mathbb{R}$. Prove that $\lim_{x\to\infty}f(x)=0$.

(f) Let $f:(1,\infty)\to\mathbb{R}$ and define $g:(0,1)\to\mathbb{R}$ by $g(x)=f\left(\frac{1}{x}\right)$ for all $x\in(0,1)$. Prove that $\lim_{x\to\infty} f(x)=L$ if and only if $\lim_{x\to 0} g(x)=L,$ where L is a real number or $\pm\infty$.

(g) The one-sided limits may be defined as follows:

The **left limit** at b of a real valued function f defined on a nonempty interval (a,b) is said to be L if for every $\varepsilon>0$ there exists $\delta\in(0,b-a)$ such that

$$|f(x)-L|<\varepsilon \text{ for all } x\in(b-\delta,b).$$

In this case, we use the notation $\lim_{x\to b^-} f(x)=L$ or $f(b-)=L.$

The **right limit** at b of a real valued function f defined on a nonempty interval (a,b) is said to be L if for every $\varepsilon>0$ there exists $\delta\in(0,b-a)$ such that

$$|f(x)-L|<\varepsilon \text{ for all } x\in(a,a+\delta).$$

In this case, we use the notation $\lim_{x\to a^+} f(x)=L$ or $f(a+)=L.$

Prove that if $f:(a,b)\to\mathbb{R}$ and $c\in(a,b)$, then $\lim_{x\to c} f(x)=L$ if and only if $f(c-)=f(c+)=L.$

3.3 Continuous Functions

In this section, we define what it means for a real valued function of a real variable to be continuous at a point as well as on a subset of the real number system. We will derive some basic properties of continuous functions along with some topological characterizations.

3.3D1 Definition: Continuity at a Point and on a Set

Let $E\subseteq\mathbb{R}$ and $f:E\to\mathbb{R}$. Then f is said to be **continuous at a point** $p\in E$ if for every $\varepsilon>0$ there exists $\delta>0$ such that

$$|f(x)-f(p)|<\varepsilon \text{ whenever } x\in E \text{ and } |x-p|<\delta.$$

The function f is said to be **continuous on a subset** A **of** E if it is continuous at each point of A. f is said to be **continuous** if it is continuous at each point of its **domain** E. f is said to be **discontinuous** at p if it is not continuous at p. In this case, p is called a **discontinuity** of f.

Note: A function is continuous at each isolated point of its domain. However, if p is a limit point of E, then f is continuous at p if and only if $\lim_{x\to p} f(x)=f(p)$. Furthermore, in the latter case, f is continuous at p if and only if $\lim_{n\to\infty} f(p_n) = f(p)$ for every sequence $\{p_n\}$ in E with $\lim_{n\to\infty} p_n = p$ (3.2T1).

Note: We have already shown that if f is a polynomial, then $\lim_{x\to p} f(x) = f(p)$ for all $p \in \mathbb{R}$ (3.2E4). We have also shown that if f is a rational function, then $\lim_{x\to p} f(x) = f(p)$ for all $p \in Dom(f)$. Hence, all polynomials and all rational functions are continuous functions.

3.3E1 Example: Let $f:\mathbb{R}\to\mathbb{R}$ be defined as $f(x) = \begin{cases} 2x & \text{if } x\in\mathbb{Q} \\ 3-x & \text{if } x\notin\mathbb{Q}. \end{cases}$

We have already shown that this function has a limit only at 1 (3.2E3); $\lim_{x\to 1} f(x) = 2$. Since $f(1)=2$, f is continuous only at the point 1. The other points are discontinuities of f.

3.3E2 Example: Define $f:(0,1)\to\mathbb{R}$ as

$$f(x) = \begin{cases} \dfrac{1}{n} & \text{if } x\in\mathbb{Q} \text{ and } x = \dfrac{m}{n} \text{ in reduced form} \\ 0 & \text{if } x\notin\mathbb{Q}. \end{cases}$$

Show that f is continuous at each irrational point in $(0,1)$, but it is discontinuous at each rational point in $(0,1)$.

Since $f(x)=0$ at each irrational point, for f to be continuous at these points, we must show that $\lim_{x\to p} f(x) = 0$ for each $p\in(0,1)$. If this is the case, then f is discontinuous at each rational point in $(0,1)$. Let $\varepsilon > 0$ be given and choose $n_0 \in \mathbb{N}$ such that $n_0 > \frac{1}{\varepsilon}$. Then there exist only a finite number of rational numbers in reduced form $\frac{m}{n}$ in $(0,1)$ such that $n < n_0$. Let these numbers be denoted by $r_1, r_2, \ldots, r_k$. If $p\in(0,1)$, define

$$\delta = \min\{|r_i - p| \mid i = 1,2,\ldots,k;\ r_i \neq p\}.$$

Then $\delta > 0$ because $r_i \neq p$ for $i = 1,2,\ldots,k$. Therefore,

$$|f(x) - 0| = f(x) \le \frac{1}{n} \le \frac{1}{n_0} < \varepsilon$$

whenever $x \in N_\delta'(p)\cap(0,1)$, where $x = \frac{m}{n}$ in reduced form when $x\in\mathbb{Q}$. Hence, f is continuous at $p\in(0,1)$ if p is irrational.

If p is rational, then $f(p)=\frac{1}{n}>0$ and f is discontinuous at p.

3.3T1 Theorem:

Let $E\subseteq\mathbb{R}$ and $f,g:E\to\mathbb{R}$ be continuous at $p\in E$. Then

(a) $f+g$ is continuous at p

(b) fg is continuous at p

(c) f/g is continuous at p if $g(p)\neq 0$.

Proof: If p is an isolated point of E, then all of the above functions are continuous at p. If p is a limit point, then (a), (b), and (c) follow easily from the analogous results for limits.

3.3T2: Theorem Composite Functions

Let $A,B\subseteq\mathbb{R}$ and let $f:A\to\mathbb{R}$ and $g:B\to\mathbb{R}$ be functions such that $Ran(f)\subseteq B$ If f is continuous at $p\in A$ and g is continuous at $f(p)$, then the composite function $h=g\circ f$ is continuous at p.

Proof: Let $\varepsilon>0$ be given. Then there exists $\eta>0$ such that

$$|g(y)-g(f(p))|<\varepsilon \text{ whenever } y\in B\cap N_\eta(f(p)).$$

Since f is continuous at p, there exists $\delta>0$ such that

$$|f(x)-f(p)|<\eta \text{ whenever } x\in A\cap N_\delta(p).$$

Therefore, if $x\in A\cap N_\delta(p)$, then $|f(x)-f(p)|<\eta$ and

$$|h(x)-h(p)|=|g(f(x))-g(f(p))|<\varepsilon.$$

Hence, h is continuous at p.

Note: The above result was fairly easy to establish. However, some of the important results on properties of continuous functions can easily be proved using a topological characterization of continuity.

3.3E3 Example: Let $E\subseteq\mathbb{R}$ and $f:E\to\mathbb{R}$ be continuous at $p\in E$. Show that

(a) $|f|$ is continuous at p.

(b) g is continuous at p, where $g:E\to\mathbb{R}$ is defined by $g(x)=\sqrt{|f(x)|}$ for all $x\in E$.

(c) If $n\in\mathbb{N}$, then f^n, defined by $f^n(x)=(f(x))^n$ for all $x\in E$, is continuous at p.

(d) f^+ defined by $f^+(x)=\max\{f(x),0\}$ for all $x\in E$, is continuous at p.

(a) Let $\varepsilon > 0$ be given. Then there exists $\delta > 0$ such that

$$|f(x) - f(p)| < \varepsilon \text{ for all } x \in N_\delta(p) \cap E.$$

Therefore,

$$\big||f(x)| - |f(p)|\big| \le |f(x) - f(p)| < \varepsilon \text{ for all } x \in N_\delta(p) \cap E.$$

Hence, $|f|$ is continuous at *p*.

(b) We shall first show that $h:[0,\infty] \to \mathbb{R}$ defined by $h(x) = \sqrt{x}$ is continuous. Suppose $p = 0$. Let $\varepsilon > 0$ be given and $\delta = \varepsilon^2$. Then

$$|h(x) - 0| = |\sqrt{x}| < \sqrt{\delta} = \sqrt{\varepsilon^2} = \varepsilon \text{ whenever } 0 \le x < \delta.$$

Suppose $p > 0$. Let $|x - p| < \frac{p}{2}$. Then $\frac{p}{4} < \frac{p}{2} < x < \frac{3p}{2}$ and

$$|h(x) - h(p)| = |\sqrt{x} - \sqrt{p}| = \frac{|x - p|}{\sqrt{x} + \sqrt{p}}$$

$$< \frac{|x - p|}{\sqrt{p/2} + \sqrt{p}} = \frac{2|x - p|}{3\sqrt{p}}.$$

Let $\varepsilon > 0$ be given and $\delta = \min\left\{\frac{p}{2}, \frac{3\sqrt{p\varepsilon}}{2}\right\}$. Then

$$|h(x) - h(p)| = |\sqrt{x} - \sqrt{p}| < \frac{2|x - p|}{3\sqrt{p}} < \frac{2 \cdot 3\sqrt{p\varepsilon}}{3\sqrt{p} \cdot 2} = \varepsilon$$

whenever $x \in N_\delta(p) \cap [0,\infty)$. Hence, *h* is continuous. Since $g = h \circ |f|$, *f* is continuous at *p*, and *h* is continuous at $|f(p)|$, *g* is continuous at *p* by (3.2T2).

(c) Define $g:\mathbb{R} \to \mathbb{R}$ by $g(x) = x^n, x \in \mathbb{R}$. Then $f^n = g \circ f$ and since *f* is continuous at *p*, and *g* is continuous at $f(p)$, f^n is continuous at *p* by (3.2T2).

(d) We can write $f^+ = \frac{1}{2}(f + |f|)$. Since *f* and $|f|$ are continuous at *p*, f^+ is continuous at *p* by (3.2T1(a),(b)).

3.3T3 Theorem: Topological Definition of Continuity

Let $E \subseteq \mathbb{R}$, $f : E \to \mathbb{R}$. Then *f* is continuous if and only if $f^{-1}(V)$ is open in *E* for every open subset *V* of $\mathbb{R}$.

Proof: Suppose f is continuous and V is an open subset of $\mathbb{R}$. If $f^{-1}(V)=\phi$, then $f^{-1}(V)$ is open in E. Suppose $p\in f^{-1}(V)$. Then $f(p)\in V$. Since V is open, there exists $\varepsilon>0$ such that $N_\varepsilon(f(p))\subseteq V$. But f is continuous at p. Therefore, there exists $\delta>0$ such that

$$f(x)\in N_\varepsilon(f(p)) \text{ for all } x\in N_\delta(p)\cap E,$$

i.e., $N_\delta(p)\cap E\subseteq f^{-1}(V)$. Since $p\in f^{-1}(V)$ was arbitrary, $f^{-1}(V)$ is open in E.

Conversely, suppose that $f^{-1}(V)$ is open in E for every open subset V of $\mathbb{R}$. Let $p\in E$ and $\varepsilon>0$ be given. Then $N_\varepsilon(f(p))$ is open and $f^{-1}(N_\varepsilon(f(p)))$ is open in E. Therefore, there exists $\delta>0$ such that

$$N_\delta(p)\cap E\subseteq f^{-1}(N_\varepsilon(f(p))),$$

i.e., $f(x)\in N_\varepsilon(f(p))$ for all $x\in N_\delta(p)\cap E$. Therefore, f is continuous at p. Hence f is continuous.

3.3T3C Corollary:

Let $E\subseteq\mathbb{R}$, $f:E\to\mathbb{R}$. Then f is continuous if and only if $f^{-1}(V)$ is closed in E for every closed subset V of $\mathbb{R}$.

Proof: This follows easily from the above theorem because $f^{-1}(V^c)=(f^{-1}(V))^c$ for any $V\subseteq\mathbb{R}$ and U is closed in E if and only if U^c is open in E.

3.3T4 Theorem:

If K is a compact subset of $\mathbb{R}$ and $f:K\to\mathbb{R}$ is continuous, then $f(K)$ is compact.

Proof: Let $\{V_\alpha\}_{\alpha\in A}$ is an open cover of $f(K)$. By theorem (3.3T3), $f^{-1}(V_\alpha)$ is open in K for every $\alpha\in A$. By theorem (3.1T8), for each $\alpha\in A$ there exists an open subset U_α of $\mathbb{R}$ such that $f^{-1}(V_\alpha)=K\cap U_\alpha$. We will now show that $\{U_\alpha\}_{\alpha\in A}$ is an open cover of K. Suppose $p\in K$. Then $f(p)\in f(K)$ and therefore, $f(p)\in V_\alpha$ for some $\alpha\in A$. Thus, $p\in f^{-1}(V_\alpha)$ and must also be in U_α. Hence, the collection $\{U_\alpha\}_{\alpha\in A}$ is an open cover of K. Since K is compact, there exist $\alpha_1, \alpha_2,..., \alpha_n\in A$ such that $K\subseteq\bigcup_{k=1}^{n}U_{\alpha_k}$. Thus

$$K=\bigcup_{k=1}^{n}U_{\alpha_k}\cap K=\bigcup_{k=1}^{n}f^{-1}(V_{\alpha_k}).$$

It follows that $f(K)=\bigcup_{k=1}^{n}f(f^{-1}(V_{\alpha_k}))$. Since $f(f^{-1}(V))\subseteq V$, $f(K)\subseteq$ V Hence, $f(K)$ is compact.

3.3T4C Corollary:

Let K be a compact subset of $\mathbb{R}$ and let $f: K \to \mathbb{R}$ be continuous. Then there exist $p, q \in K$ such that $f(q) \leq f(x) \leq f(p)$ for all $x \in K$.

Proof: By the above theorem, $f(K)$ is compact and must be closed and bounded (3.1T10). Therefore, it has a finite supremum, say M. Since $f(K)$ is closed, $M \in f(K)$, i.e., there exists $p \in K$ such that $f(p) = M$. Similarly, there exists $q \in K$ such that $f(q) = m = \inf f(K)$. Thus, $m = f(q) \leq f(x) \leq f(p) = M$ for all $x \in K$.

Note: If the domain of f is not compact, then the above conclusion cannot be made, e.g., the function $f:(0,1] \to \mathbb{R}$ defined by $f(x) = \frac{1}{x}$ is unbounded and $\sup f((0,1]) = \infty$ and the function $g:[1,\infty) \to \mathbb{R}$ defined by $g(x) = \frac{1}{x}$ is unbounded and $\inf g([1,\infty)) = 0 \notin g([1,\infty))$.

3.3T5 Theorem: Intermediate Value Theorem

Let $f:[a,b] \to \mathbb{R}$ be continuous, where $a < b$. If $\gamma \in \mathbb{R}$ is between $f(a)$ and $f(b)$, then there exists $c \in [a,b]$ such that $f(c) = \gamma$.

Proof: If $f(a) = f(b)$, then $\gamma = f(a) = f(b)$, i.e., $c = a$. Suppose $f(a) \neq f(b)$. If $\gamma = f(a)$, then $c = a$. If $\gamma = f(b)$, then $c = b$. Suppose γ is strictly between $f(a)$ and $f(b)$. Without loss of generality, we may assume that $f(a) < f(b)$. In this case, $f(a) < \gamma < f(b)$. Let

$$A = \{x \in [a,b] \,|\, f(x) \leq \gamma\}.$$

Then $A \neq \phi$ because $a \in A$, and A is bounded above by b. Therefore,

$$\sup A = c \text{ exists in } \mathbb{R}.$$

Since b is an upper bound of A, $c \leq b$. Since $c = \sup A$, either $c \in A$ or $c \in A'$. If $c \in A$, then $f(c) \leq \gamma$. If $c \in A'$, then by (2.4T2(b)), there exists a sequence $\{p_n\}$ in E such that $\lim_{n\to\infty} p_n = c$. For all $n \in \mathbb{N}$, $f(p_n) \leq \gamma$ since $p_n \in A$. Therefore,

$$f(c) = \lim_{n\to\infty} f(p_n) \leq \gamma$$

because f is continuous. In any case, $f(c) \leq \gamma$.

Suppose $f(c) < \gamma$. Let $\varepsilon = \frac{1}{2}(\gamma - f(c)) > 0$. Since f is continuous at c, there exists $\delta > 0$ such that

$$f(c) - \varepsilon < f(x) < f(c) + \varepsilon \text{ whenever } x \in N_\delta(c) \cap [a,b].$$

Since $f(c)<\gamma<f(b)$, $c \neq b \Rightarrow c<b$. Thus $(c,b)\cap N_\delta(c)\neq \phi$. Let $x\in(c,\min\{b,c+\delta\})$. Then

$$f(x)<f(c)+\varepsilon=f(c)+\frac{1}{2}(\gamma-f(c))$$
$$=\frac{1}{2}(\gamma+f(c))<\gamma.$$

This implies that $x\in A$ and $x>c$, which contradicts $c=\sup A$. Hence, $f(c)=\gamma$.

Note: If γ is strictly between $f(a)$ and $f(b)$, then $a<c<b$.

3.3T5C Corollary:

Suppose $f:I\to\mathbb{R}$ is continuous, where *I* is an interval in $\mathbb{R}$. Then $f(I)$ is also an interval in $\mathbb{R}$.

Proof: Suppose $\alpha,\beta\in f(I)$ and $\alpha<\gamma<\beta$. Then there exists $a,b\in I$ such that

$$f(a)=\alpha \text{ and } f(b)=\beta.$$

Since $\alpha<\beta$, $a\neq b$. Let $c=\min\{a,b\}$ and $d=\max\{a,b\}$. Then $c<d$. Since *I* is an interval, $[c,d]\subseteq I$ and *f* is continuous on $[c,d]$. By the above theorem, there exists $\xi\in[c,d]$ such that

$$f(\xi)=\gamma.$$

Thus, $\gamma\in f(I)$ and $f(I)$ is an interval in $\mathbb{R}$.

3.3E4 Example: Let $f:[a,b]\to[a,b]$ be continuous, where $a<b$. Show that there exists $\xi\in[a,b]$ such that $f(\xi)=\xi$.

Define $g(x)=x-f(x)$ for all $x\in[a,b]$. Then *g* is clearly continuous on $[a,b]$,

$$g(a)=a-f(a)\leq 0,\ g(b)=b-g(b)\geq 0.$$

Therefore, by (3.3T4), there exists $\xi\in[a,b]$ such that $g(\xi)=0$. Hence, $f(\xi)=\xi$.

3.3T6 Theorem:

Suppose $x>0$ and $n\in\mathbb{N}$. Then there exists a unique positive real number *y* such that $y^n=x$. This number *y* is called the nth root of *x* and is denoted by $\sqrt[n]{x}$ or $x^{1/n}$.

Proof: Let $f:\mathbb{R}\to\mathbb{R}$ be defined by $f(x)=x^n$. We have already established that *f* is continuous. Since *f* is continuous on $[0,x+1]$ and $f(0)=0<x<(x+1)^n=f(x+1)$, there exists $y\in(0,x+1)$ such that $y^n=f(y)=x$. This proves the existence of such a number *y*.

Suppose $0<y_1<y_2$. Then

$$0 < y_1^2 < y_1 y_2 < y_2^2.$$

It follows that $0 < y_1^n < y_2^n$. Hence, $y_1, y_2 > 0$ and

$$y_1 \neq y_2 \Rightarrow y_1^n \neq y_2^n.$$

Therefore,

$$y_1, y_2 > 0 \text{ and } y_1^n = y_2^n \Rightarrow y_1 = y_2.$$

This proves the uniqueness of *y*.

3.3E5 Example: Let $f : \mathbb{R} \to \mathbb{R}$ and let *E* be a dense subset of $\mathbb{R}$. Prove that if *f* is continuous and $f(x) = p \in \mathbb{R}$ for all $x \in E$, then $f(x) = p$ for all $x \in \mathbb{R}$.

Suppose $f(y) = q \neq p$ for some $y \in \mathbb{R}$. Then $\varepsilon = |p - q| > 0$ and there exists $\delta > 0$ such that

$$|f(x) - f(y)| < \varepsilon \text{ whenever } x \in N_\delta(y).$$

Since *E* is dense, there exists $x_0 \in E \cap N_\delta(y)$. Then

$$|p - q| = |f(x_0) - f(y)| < \varepsilon = |p - q|.$$

This is a contradiction. Hence, result.

3.3A1 Assignment:

(a) Suppose $f : \mathbb{R} \to \mathbb{R}$ is continuous and $f(0) = f(2)$. Prove that there exists $c \in [0,1]$ such that $f(c) = f(c+1)$.

(b) Suppose $f : \mathbb{R} \to \mathbb{R}$ satisfies $f(x+y) = f(x) + f(y)$ for all $x, y \in \mathbb{R}$ and is continuous at a point $p \in \mathbb{R}$. Prove that *f* is continuous and $f(x) = cx$ for all $x \in \mathbb{R}$, where $c = f(1)$.

(c) Suppose $f, g : E \to \mathbb{R}$ are continuous functions, where $E \subseteq \mathbb{R}$. Show that the set $\{x \in E \mid f(x) < g(x)\}$ is open in *E*.

(d) Suppose $f : K \to \mathbb{R}$ is one-to-one and continuous on *K*, where $K \subseteq \mathbb{R}$ is compact. Deduce that $f(K)$ is compact and use (3.3T3) to show that f^{-1} is a continuous one-to-one function from $f(K)$ onto *K*.

(e) Suppose $f : \mathbb{R} \to \mathbb{R}$ is continuous and satisfies $f(0) = 1$ and $f(x+y) = f(x)f(y)$ for all $x, y \in \mathbb{R}$. Prove that $f(x) = a^x$ for all $x \in \mathbb{R}$, where $a = f(1)$.

3.4 Uniform Continuity

A function $f: E \to \mathbb{R}$ is continuous on E if it is continuous at each $p \in E$. In the $\varepsilon - \delta$ definition of continuity $\delta = \delta(\varepsilon, p)$. However, there may be functions that satisfy the $\varepsilon - \delta$ definition of continuity with $\delta = \delta(\varepsilon)$, i.e., δ is independent of p.

3.4D1 Definition: Uniform Continuity

Let $f: E \to \mathbb{R}$, where $E \subseteq \mathbb{R}$. Then f is said to be **uniformly continuous on *E*** if for every $\varepsilon > 0$ there exists $\delta > 0$ such that

$$|f(x) - f(y)| < \varepsilon \text{ whenever } x, y \in E \text{ and } |x - y| < \delta.$$

3.4E1 Example: Let $f: \mathbb{R} \to \mathbb{R}$ be defined by $f(x) = x^2$ for all $x \in \mathbb{R}$.

(i) Suppose E is a bounded subset of real numbers. Then

$|x| \le M$ for all $x \in E$ where $M > 0$. Therefore, for all $x, y \in E$,

$$\begin{aligned} |f(x) - f(y)| &= |x^2 - y^2| = |x + y||x - y| \\ &= \big||x| + |y|\big||x - y| \\ &\le 2M|x - y|. \end{aligned}$$

Let $\varepsilon > 0$ be given and let $\delta = \dfrac{\varepsilon}{2M}$. Then for all $x, y \in E$ with $|x - y| < \delta$,

$$\begin{aligned} |f(x) - f(y)| &\le 2M|x - y| \\ &< 2M\frac{\varepsilon}{2M} \\ &= \varepsilon. \end{aligned}$$

Therefore, f is uniformly continuous on E.

(ii) Suppose f is uniformly continuous on $[a, \infty)$ where $a \in \mathbb{R}$. Then there exists $\delta > 0$ such that

$$|f(x) - f(y)| < 1 \text{ whenever } x, y \in [a, \infty) \text{ and } |x - y| < \delta.$$

Let $\gamma = \max\{\delta^{-1}, |a|\}$ and let $x = \gamma$ and $y = \gamma + \gamma^{-1}/2$. Then $x, y \in [a, \infty)$ and $|x - y| = \gamma^{-1}/2 \le \delta/2 < \delta$, but

$$\begin{aligned} |f(x) - f(y)| &= |x^2 - y^2| \\ &= \left|(\gamma + \gamma^{-1}/2)^2 - \gamma^2\right| \\ &= 1 + \gamma^{-2}/4 > 1. \end{aligned}$$

This is a contradiction. Therefore, *f* is not uniformly continuous on $[a,\infty)$.

3.4A1 Assignment:

Show that the function $f:(0,\infty)\to\mathbb{R}$ defined by $f(x)=\frac{1}{x^2}$ for all $x\in(0,\infty)$ is not uniformly continuous on $(0,\infty)$ but is uniformly continuous on $[a,\infty)$ with $a>0$.

3.4E2 Example: Consider the function $f:(0,\infty)\to\mathbb{R}$ defined by $f(x)=\frac{1}{x}$ for all $x\in(0,\infty)$.

(i) Suppose *f* is uniformly continuous on $(0,1]$. Then there exists $\delta>0$ such that

$$|f(x)-f(y)|<1 \text{ for all } x,y\in(0,1] \text{ and } |x-y|<\delta.$$

Let $\gamma=\min\{\delta,1\}$, $x=\gamma/2$, and $y=\gamma$. Then $x,y\in(0,1]$ and

$$|x-y|=\gamma/2\le\delta/2<\delta,$$

but

$$|f(x)-f(y)|=\left|\frac{2}{\gamma}-\frac{1}{\gamma}\right|=\left|\frac{1}{\gamma}\right|\ge 1.$$

This is a contradiction. Therefore, *f* is not uniformly continuous on $(0,1]$.

(ii) Suppose $E=[a,\infty)$ where $a>0$. Then for all $x,y\in E$,

$$|f(x)-f(y)|=\left|\frac{1}{x}-\frac{1}{y}\right|=\left|\frac{y-x}{xy}\right|\le\frac{1}{a^2}|x-y|.$$

Let $\varepsilon>0$ be given and let $\delta=a^2\varepsilon$. Then for all $x,y\in E$ with $|x-y|<\delta$,

$$|f(x)-f(y)|\le\frac{1}{a^2}|x-y|<\frac{1}{a^2}a^2\varepsilon=\varepsilon.$$

Thus, *f* is uniformly continuous on *E*.

3.4D2 Definition: Lipschitz Functions

Let $f:E\to\mathbb{R}$, where $E\subseteq\mathbb{R}$. Then *f* satisfies a **Lipschitz condition on *E*** if there exists $M>0$ such that $|f(x)-f(y)|\le M|x-y|$ for all $x,y\in E$. In this case, *f* is called a **Lipschitz Function.** The constant *M* is sometimes referred to as a Lipschitz constant.

3.4T1 Theorem:

If *f* is a Lipschitz function defined on $E\subseteq\mathbb{R}$, then *f* is uniformly continuous on *E*.

Proof: Let $\varepsilon>0$ be given and $\delta=\varepsilon/M$. Then for all $x,y\in E$,

$$\left|f(x)-f(y)\right| \leq M\left|x-y\right| < M\varepsilon / M = \varepsilon$$

whenever $\left|x-y\right| < \delta$.

3.4T2 Theorem:

Suppose $f: K \to \mathbb{R}$ is continuous, where *K* is a compact subset of $\mathbb{R}$. Then f is uniformly continuous on *K*.

Proof: Let $\varepsilon > 0$ be given. Then for each $p \in K$ there exists $\delta(p) > 0$ such that for all $x \in K$ with $\left|x-p\right| < 2\delta(p)$,

$$\left|f(x)-f(p)\right| < \varepsilon / 2.$$

The collection $\left\{N_{\delta(p)}(p) \middle| p \in K\right\}$ is an open cover of the compact set *K* and must have a finite subcover of *K*. Hence, there exists a finite number of points $p_1, p_2, \ldots, p_n \in K$ such that $K \subseteq \bigcup_{k=1}^{n} N_{\delta(p_k)}(p_k)$. Let $\delta = \min\left\{\delta(p_k) \middle| k = 1, 2, \ldots, n\right\}$. Suppose $x, y \in K$ and $\left|x-y\right| < \delta$. Then $x \in N_{\delta(p_k)}(p_k)$ for some *k*. Since $\left|x-y\right| < \delta \leq \delta(p_k)$, $x, y \in N_{2\delta(p_k)}(p_k)$. Therefore,

$$\begin{aligned}\left|f(x)-f(y)\right| &\leq \left|f(x)-f(p_k)\right| + \left|f(p_k)-f(y)\right| \\ &< \frac{\varepsilon}{2} + \frac{\varepsilon}{2} = \varepsilon.\end{aligned}$$

Hence, *f* is uniformly continuous on *K*.

3.4T2C Corollary:

A continuous real valued function defined on a closed and bounded interval of the form $[a,b]$ where $-\infty < a < b < \infty$, is uniformly continuous on $[a,b]$.

Proof: Follows immediately from the above theorem.

Note: In (3.4E1), *f* is clearly continuous on $[a,\infty)$ but is not uniformly continuous there. This shows that boundedness of *E* is necessary. In (3.4E2), *f* is clearly continuous on $(0,1]$ but is not uniformly continuous there. This shows that the closure of *E* is also necessary.

3.4A2 Assignment:

(a) Suppose $f, g: E \to \mathbb{R}$ are uniformly continuous functions, where $E \subseteq \mathbb{R}$.

- (i) Prove that $f+g$ is uniformly continuous on *E*.
- (ii) If, in addition *f* and *g* are bounded, prove that *fg* is uniformly continuous on *E*.
- (iii) Is (ii) true if only one of the functions is bounded? Justify your answer by proving the statement or providing a counterexample.

(b) Suppose $f : E \to \mathbb{R}$ is a uniformly continuous function where $E \subseteq \mathbb{R}$.

(i) If $\{x_n\}$ is a Cauchy sequence in *E*, prove that $\{f(x_n)\}$ is also a Cauchy sequence.

(ii) If $E = (a,b)$ where $a < b$, use (i) to define *f* at the end points *a* and *b* so that *f* is continuous on $[a,b]$.

(iii) If *E* is bounded, prove that *f* is bounded on *E*.

(c) Suppose $f : [a,\infty) \to \mathbb{R}$ is a continuous function where $a \in \mathbb{R}$ and $\lim_{x\to\infty} f(x) = L \in \mathbb{R}$. Prove that *f* is bounded on $[a,\infty)$ and *f* is uniformly continuous on $[a,\infty)$.

3.5 Discontinuities and Monotonic Functions

A real valued function fails to be continuous at a point $p \in \mathbb{R}$ if *p* is not in its domain, *p* is in its domain but $\lim_{x\to p} f(x)$ does not exist, or $f(p) \neq \lim_{x\to p} f(x)$. In the latter case, we could redefine *f* at *p* to have equality. In order to analyze this further, we need to consider one-sided limits of functions.

3.5D1 Definition: Left and Right Limits of a Function at a Point

Let $f : E \to \mathbb{R}$, where $E \subseteq \mathbb{R}$ and suppose *p* is a limit point of $E \cap (p,\infty)$. Then *f* has a **right limit at *p*** if there exists $L \in \mathbb{R}$ such that for every $\varepsilon > 0$ there exists $\delta > 0$ such that

$$|f(x) - L| < \varepsilon \text{ for all } x \in E \cap (p, p+\delta).$$

In this case, *L* is called the right limit of *f* at *p* and this limit is denoted by

$$\lim_{x\to p+} f(x) \text{ or } f(p+).$$

Similarly, if *p* is a limit point of $E \cap (-\infty, p)$, the **left limit of *f* at *p*** is defined to be the number $L \in \mathbb{R}$, if for every $\varepsilon > 0$ there exists $\delta > 0$ such that

$$|f(x) - L| < \varepsilon \text{ for all } x \in E \cap (p-\delta, p).$$

The left limit is denoted by

$$\lim_{x\to p-} f(x) \text{ or } f(p-).$$

3.5T1 Theorem:

Let $f : E \to \mathbb{R}$, where $E \subseteq \mathbb{R}$ and suppose *p* is a limit point of both $E \cap (-\infty, p)$ and $E \cap (p,\infty)$. Then $\lim_{x\to p} f(x)$ exists if and only if both $\lim_{x\to p-} f(x)$ and $\lim_{x\to p+} f(x)$ exist and $\lim_{x\to p-} f(x) = \lim_{x\to p+} f(x) = L \in \mathbb{R}$. In this case, $\lim_{x\to p} f(x) = L$.

Proof: Suppose $\lim_{x\to p-} f(x)$ and $\lim_{x\to p+} f(x)$ exist and $\lim_{x\to p-} f(x)=\lim_{x\to p+} f(x)=L\in\mathbb{R}$. Let $\varepsilon>0$ be given. Then there exists $\delta_1,\delta_2>0$ such that

$$|f(x)-L|<\varepsilon \text{ whenever } x\in E\cap(p-\delta_1,p)$$

and

$$|f(x)-L|<\varepsilon \text{ whenever } x\in E\cap(p,p+\delta_2).$$

Let $\delta=\min\{\delta_1,\delta_2\}$. Then

$$|f(x)-L|<\varepsilon \text{ whenever } x\in E\cap N'_\delta(p).$$

Thus, $\lim_{x\to p} f(x)=L$.

Conversely, suppose $\lim_{x\to p} f(x)=L$. If $\varepsilon>0$ is given, Then there exists $\delta>0$ such that

$$|f(x)-L|<\varepsilon \text{ whenever } x\in E\cap N'_\delta(p).$$

Thus,

$$|f(x)-L|<\varepsilon \text{ whenever } x\in E\cap(p-\delta,p)$$

and

$$|f(x)-L|<\varepsilon \text{ whenever } x\in E\cap(p,p+\delta).$$

Hence, $\lim_{x\to p-} f(x)=\lim_{x\to p+} f(x)=L$.

3.5D2 Definition:

Let $f:E\to\mathbb{R}$, where $E\subseteq\mathbb{R}$ and let $p\in E$. Then f is said to be **right continuous** at p if for $\varepsilon>0$ there exists $\delta>0$ such that $|f(x)-f(p)|<\varepsilon$ whenever $x\in E\cap[p,p+\delta)$. f is said to be **left continuous** at p if for $\varepsilon>0$ there exists $\delta>0$ such that $|f(x)-f(p)|<\varepsilon$ whenever $x\in E\cap(p-\delta,p]$.

Note: If p is an isolated of E or if p is not a limit point of $E\cap(p,\infty)$, then there exists $\delta>0$ such that $E\cap(p,p+\delta)=\phi$. Thus if $\varepsilon>0$ is arbitrary, then for this δ, $|f(x)-f(p)|<\varepsilon$ for all $x\in E\cap[p,p+\delta)$. Therefore, f is right continuous at p. If $E=[a,b]$, then f is right continuous at a and f is left continuous at a if and only if it is continuous at a.

If p is an isolated of E or if p is not a limit point of $E\cap(-\infty,p)$, then there exists $\delta>0$ such that $E\cap(p-\delta,p)=\phi$. Thus if $\varepsilon>0$ is arbitrary, then for this δ,

$|f(x)-f(p)|<\varepsilon$ for all $x \in E \cap (p-\delta, p]$. Therefore, f is left continuous at p. If $E=[a,b]$, then f is left continuous at a and f is right continuous at a if and only if it is continuous at a.

3.5T2 Theorem:

Let $f:(a,b) \to \mathbb{R}$, where $a<b$ and let $p \in (a,b)$. Then f is right continuous at p if and only if $f(p+)=f(p)$ and f is left continuous at p if and only if $f(p-)=f(p)$.

Proof: Follows immediately from (3.5D1) and (3.5D2).

3.5T3 Theorem:

Let $f: E \to \mathbb{R}$, where $E \subseteq \mathbb{R}$ and let $p \in E$. Then f is continuous at p if and only if it is both right continuous and left continuous at p.

Proof: If p is an isolated point of E, then the result holds. Let p be a limit point of $E \cap (p, \infty)$ but not a limit point of $E \cap (-\infty, p)$. Then f is left continuous at p. Therefore, f is continuous at p if and only if

$$f(p)=\lim_{x \to p} f(x)=f(p+),$$

i.e., f is right continuous at p and left continuous at p.

Similarly, the result holds if p is a limit point of $E \cap (-\infty, p)$ but not a limit point of $E \cap (p, \infty)$.

If p is a limit point of both $E \cap (-\infty, p)$ and $E \cap (p, \infty)$, then f is continuous at p if and only if

$$f(p)=\lim_{x \to p} f(x)=f(p+)=f(p-),$$

i.e., f is both left continuous and right continuous at p.

Note: A function $f:(a,b) \to \mathbb{R}$, where $a<b$ is continuous at $p \in (a,b)$ if and only if $f(p-)$ and $f(p+)$ exist and $f(p)=f(p-)=f(p+)$. Therefore, a function f, defined on an interval I can fail to be continuous at $p \in \bar{I}$, for several reasons. One possibility is that $\lim_{x \to p} f(x)$ exists but is not equal to $f(p)$ for $p \in I$. Another possibility is that $f(p-) \neq f(p+)$. A third possibility is that $f(p-)$ or $f(p+)$ might not exist.

3.5D3 Definition: Types of Discontinuities

Let $f:(a,b) \to \mathbb{R}$, where $a<b$. Then

(a) *f* is said to have a discontinuity of the **first kind** or a **simple** discontinuity at *p* if *p* is a discontinuity of *f* and both $f(p-)$ and $f(p+)$ exist. This discontinuity is called a **removable discontinuity** if $f(p-)=f(p+)$; else it is called a **jump discontinuity**.

(b) *f* is said to have a discontinuity of the **second kind** if $f(p-)$ or $f(p+)$ does not exist.

3.5E1 Example: Consider the function $f:(0,\infty)\to\mathbb{R}$ defined by

$$f(x)=\begin{cases} 2x+1, & 0<x\le 1 \\ \dfrac{1}{x-1}, & 1<x<2 \\ 2x-3, & 2\le x<3 \\ 3x-1, & 3\le x \end{cases}$$

It is easily seen that $f(0+)=1$, $f(1-)=3$, does not exist, $f(2-)=f(2)=f(2+)=1$, $f(3-)=3$, and $f(3+)=7$. Therefore, *f* has discontinuities at 0, 1, and 3. It is continuous at all other points of its domain. *f* has a removable discontinuity at 0, a discontinuity of the second kind at 1, and a jump discontinuity at 3. Note: 0 is a limit point of the domain of *f*.

3.5E2 Example: The greatest integer function is defined as follows:

$$f(x)=[x]=n \text{ if } n\le x<n+1, \text{ where } n\in\mathbb{Z}.$$

For example, $f(3.5)=3$, $f(-5)=-5$, and $f(-9.3)-10$. It follows that *f* is continuous at each point in $\mathbb{R}\setminus\mathbb{Z}$ and has a jump discontinuity at each point in $\mathbb{Z}$, because $f(n-)=n-1$ and $f(n+)=n=f(n)$ for each $n\in\mathbb{Z}$ and $f(x-)=f(x)=f(x+)$ for each $x\in\mathbb{R}\setminus\mathbb{Z}$.

Note: In (3.4E1), the function *f* is clearly unbounded and its value at points close to 1 and to the right of 1 can be made as large as we like by taking *x* sufficiently close to 1. Therefore, we now formally define the one-sided infinite limits of a function.

3.5D4 Definition: One-sided Infinite Limits

Let $f:E\to\mathbb{R}$, where $E\subseteq\mathbb{R}$ and suppose *p* is a limit point of $E\cap(p,\infty)$. Then *f* is said to have the right limit ∞ if for every $X>0$ there exists $\delta>0$ such that

$$f(x)>X \text{ whenever } x\in E\cap(p,p+\delta).$$

In this case, we write

$$\lim_{x\to p+} f(x)=f(p+)=\infty.$$

The function *f* is said to have the right limit $-\infty$ if for every $X>0$ there exists $\delta>0$ such that

$$f(x)<-X \text{ whenever } x\in E\cap(p,p+\delta).$$

In this case, we write

$$\lim_{x \to p+} f(x) = f(p+) = -\infty.$$

In a similar manner, we can define

$$\lim_{x \to p-} f(x) = f(p-) = \infty \text{ and } \lim_{x \to p-} f(x) = f(p-) = -\infty$$

by considering *p* to be a limit point of $E \cap (-\infty, p)$.

Note: In (3.4E1), $f(1+) = \lim_{x \to 1+} f(x) = \infty.$

3.5D5 Definition: Monotonic Functions

Let $f : I \to \mathbb{R}$ where *I* is a nonempty interval with at least two points.

(a) *f* is said to be (monotonically) **increasing** on *I* if $f(x_1) \leq f(x_2)$ for all $x_1, x_2 \in I$ with $x_1 < x_2$.

(b) *f* is said to be (monotonically) **decreasing** on *I* if $f(x_1) \geq f(x_2)$ for all $x_1, x_2 \in I$ with $x_1 < x_2$.

(c) *f* is said to be **monotonic** on *I* if *f* is increasing on *I* or decreasing on *I*.

Note: If the inequality $\leq$ in (a) is replaced by $<$, then we say that *f* is **strictly increasing** on *I*. If the inequality $\geq$ in (b) is replaced by $>$, then we say that *f* is **strictly decreasing** on *I*. The function *f* is said to be **strictly monotonic** on *I* if *f* is strictly increasing on *I* or strictly decreasing on *I*.

3.5T4 Theorem:

Let $f : I \to \mathbb{R}$ where *I* is a nonempty interval in $\mathbb{R}$. If *f* is increasing on *I*, then $f(p-)$ and $f(p+)$ exist for each $p \in I$ and

$$\sup_{x<p} f(x) = f(p-) \leq f(p) \leq f(p+) = \inf_{p<x} f(x).$$

Furthermore, if $p, q \in I$ with $p < q$, then $f(p+) \leq f(q-)$. Similar results hold for decreasing functions.

Proof: Let $p \in I$ and $E_p = \{ f(x) | x < p, \, x \in I \}$. Since *f* is increasing on *I*, E_p is bounded above by $f(p)$ and must have a real supremum, say *M*. Then $M \leq f(p)$. We must now show that $M = f(p-)$. Let $\varepsilon > 0$ be given. Then there exists $\delta > 0$ such that $(p - \delta) \in I$ and

$$M - \varepsilon < f(p - \delta) \leq M.$$

Thus, if $p - \delta < x < p$, then

$$M - \varepsilon < f(p - \delta) \leq f(x) \leq M < M + \varepsilon,$$

since *f* is increasing. Therefore, $M = f(p-)$. In a similar manner, we can prove that

$$f(p) < f(p+) = \inf\{f(x) | p < x,\ x \in I\}.$$

Suppose $p, q \in I$ with $p < q$. Then

$$f(p+) = \inf_{p<x,\ x\in I} f(x) \le \inf_{p<x<q} f(x)$$

$$\le \sup_{p<x<q} f(x) \le \sup_{x<q,\ x\in I} f(x) = f(q-).$$

Note: Monotonic functions do not have discontinuities of the second kind.

3.5T4C Corollary:

Let $f : I \to \mathbb{R}$ be monotonic, where *I* is a nonempty interval in $\mathbb{R}$. Then the set of discontinuities of *f* is at most countable.

Proof: Without loss of generality, we may assume that *f* is increasing. Let *E* be the set of discontinuities of *f*. If $p \in E$, then $f(p-) < f(p+)$. Choose $r_p \in \mathbb{Q} \cap (f(p-), f(p+))$ and define $g(p) = r_p$. Then $g : E \to \mathbb{Q}$ and

$$p < q \Rightarrow g(p) = r_p < f(p+) \le f(q-) < r_q = g(q).$$

Thus, $p \ne q \Rightarrow g(p) \ne g(q)$ and *g* is one-to-one. Therefore, *E* is equivalent to a subset of $\mathbb{Q}$ and must be at most countable.

3.5D5 Definition: Unit Step Function

The unit step function $U_a : \mathbb{R} \to \mathbb{R}$ is defined by:

$$U_a(x) = \begin{cases} 0, & x < a \\ 1, & x \ge a \end{cases}$$

where $a \in \mathbb{R}$. If we define $U = U_0$, then

$$U_a(x) = U_0(x - a) = U(x - a) \text{ for all } x \in \mathbb{R}.$$

3.5E3 Example: Let $a < b$, $A = \{a_1, a_2, \ldots, a_n\} \subseteq (a, b)$, and let $\{c_1, c_2, \ldots, c_n\}$ be a set of positive real numbers, where $n \in \mathbb{N}$. Define $f : [a, b] \to \mathbb{R}$ by

$$f(x) = \sum_{k=1}^{n} c_k U_{a_k}(x),\ x \in [a, b].$$

Since each U_{a_k} is increasing for each *k*, the function *f* defined above should also be increasing because each c_k is positive. For each *k*, U_{a_k} is continuous at all points of $[a, b]$ except a_k where it is right continuous with $U_{a_k}(a_k+) - U_{a_k}(a_k-) = 1$. Therefore, it follows that the function f is increasing on $[a,b]$, continuous on $[a,b]\setminus\{a_1,a_2,...,a_n\}$, right continuous at each point in $\{a_1,a_2,...,a_n\}$ and satisfies

$$f(a_k+)-f(a_k-)=c_k \text{ for each } k=1,\ 2,...,\ n.$$

3.5A1 Assignment:

(a) Let $f:I\to\mathbb{R}$ be one-to-one and continuous on I, where I is a nonempty interval in $\mathbb{R}$. Prove that f is strictly monotonic on I.

(b) Let $a<b$, $\{a_n\}_{n\in\mathbb{N}}\subseteq(a,b)$ with $a_m\neq a_n$ for $m\neq n$, and let $\{c_n\}_{n\in\mathbb{N}}$ be a real positive sequence such that $\sum_{n=1}^{\infty}c_n$ converges. Define $f:[a,b]\to\mathbb{R}$ by

$$f(x)=\sum_{k=1}^{n}c_kU_{a_k}(x), \text{for each } x\in[a,b].$$

Prove the following:

(i) $f(a)=0$ and $f(b)=\sum_{n=1}^{\infty}c_n$

(ii) f is continuous on $[a,b]\setminus\{a_n\mid n\in\mathbb{N}\}$

(iii) f is right continuous at each a_n, $n\in\mathbb{N}$, i.e., $f(a_n+)=f(a_n)$

(iv) $f(a_n)-f(a_n-)=c_n$, $n\in\mathbb{N}$.

(c) Let $f:[a,b)\to\mathbb{R}$ be continuous, where $a<b$. Prove that $f(b-)=\lim_{x-b-}f(x)$ exists if and only if f is uniformly continuous on $[a,b)$.

3.5T5 Theorem:

Let $f:I\to\mathbb{R}$ be strictly monotonic and continuous on I, where I is a nonempty interval in $\mathbb{R}$. Then f^{-1} is strictly monotonic and continuous on $J=f(I)$.

Proof: Without loss of generality, we may assume that f is strictly increasing. Since f is continuous, by (3.3T5(c)), $f(I)$ is an interval, say J. Furthermore, since f is strictly increasing on I, f is one-to-one function from I onto J. Hence, f^{-1} exists and is a one-to-one function from J onto I. Suppose $y_1,y_2\in J$ with $y_1<y_2$. Since f is strictly increasing, there exist $x_1,x_2\in I$ such that

$$x_1<x_2 \text{ and } f(x_1)=y_1,\ f(x_2)=y_2.$$

Therefore, $f^{-1}(y_1)<f^{-1}(y_2)$, i.e., f^{-1} is strictly increasing on J.

Suppose $q \in J$ with $J \cap (q, \infty) \neq \phi$, i.e., q is not the right endpoint of J (if it exists). Then there exists $p \in I$ such that $f(p) = q$ and $I \cap (p, \infty) \neq \phi$ because f is strictly increasing. Let $\varepsilon > 0$ be given and choose $\eta \in (0, \varepsilon)$ such that $(p + \eta) \in I$. Since f is strictly increasing and continuous,

$$f([p, p+\eta)) = [f(p), f(p+\eta)) = [q, q+\delta)$$

where $\delta = f(p+\eta) - f(p) > 0$. Since f^{-1} is strictly increasing,

$$p = f^{-1}(q) < f^{-1}(y) < f^{-1}(q+\delta) = p + \eta \text{ for all } y \in (q, q+\delta).$$

Thus

$$\left|f^{-1}(y) - f^{-1}(q)\right| < \eta < \varepsilon \text{ for all } y \in J \text{ with } q \leq y < q + \delta.$$

Hence, f^{-1} is right continuous at q. Similarly, we can show that f^{-1} is left continuous at q when $J \cap (-\infty, q) \neq \phi$. It follows that f^{-1} is continuous on J.

3.5E5 Example: The function $f : [0, \infty) \to \mathbb{R}$ defined by

$$f(x) = x^n, \ x \in \mathbb{R},$$

where $n \in \mathbb{N}$, is continuous and strictly increasing on $[0, \infty)$. See (3.3T6). Therefore, by (3.5T5), f^{-1} is also strictly increasing and continuous on its domain $[0, \infty)$. But

$$g(x) = f^{-1}(x) = x^{1/n} \text{ for all } x \in [0, \infty).$$

Hence, g is continuous and strictly increasing on $[0, \infty)$.

CHAPTER

4 Differentiation

4.1 The Derivative

In this section, we shall define the derivative of a function at a point on its domain as well as the derivative of a function on a subset of its domain. Furthermore, the domain of each of the functions considered is a nontrivial interval I in $\mathbb{R}$. We shall also establish the standard results for derivatives.

4.1D1 Definition: Derivative of a Function

Let I be an interval in $\mathbb{R}$, $f: I \to \mathbb{R}$, $E \subseteq I$, and $p \in I$.

(a) f is said to be **differentiable at *p*** if $\lim\limits_{x \to p} \dfrac{f(x)-f(p)}{x-p}$ exists. In this case, the above limit is denoted by $f'(p)$ and is called the **derivative of *f* at *p*.**

(b) f is said to be **differentiable on *E*** if f is differentiable at each point in E.

(c) The **derivative of *f*** is the function $f': E \to \mathbb{R}$, given by

$$f'(x) = \lim_{t \to x} \frac{f(t)-f(x)}{t-x}, \text{ for all } x \in E,$$

where $E = \{x \in I \mid f'(x) \text{ exists}\}$.

Note: If p is an endpoint of I, then the above limit is a one-sided limit. Therefore, it might be quite useful to define one-sided derivatives as well.

4.1D2 Definition: One-sided Derivatives

Let I be an interval in $\mathbb{R}$, $f: I \to \mathbb{R}$, and $p \in I$.

(a) f is said to be **left-differentiable** at p if $\lim_{x\to p-} \frac{f(x)-f(p)}{x-p}$ exists. In this case, the resulting limit is called the **left derivative** of f at p and is denoted by $f_-'(p)$.

(b) f is said to be **right-differentiable** at p if $\lim_{x\to p+} \frac{f(x)-f(p)}{x-p}$ exists. In this case, the resulting limit is called the **right derivative** of f at p and is denoted by $f_+'(p)$.

Note: If p is a left endpoint of I, then $f_-'(p)$ does not exist. If p is a right endpoint of I, then $f_+'(p)$ does not exist. Furthermore, f is differentiable at an interior point p of its domain if and only if it is right differentiable and left differentiable at p.

Note: In a lot of applications, the function f is defined on a closed interval of the form $[a,b]$ where $a<b$. In this case, $f'(a)=f_+'(a)$ and $f'(b)=f_-'(b)$.

4.1E1 Example: Let $f:\mathbb{R}\to\mathbb{R}$ be defined by $f(x)=x^n$, $x\in\mathbb{R}$, where $n\in\mathbb{N}$. Then

$$\frac{f(t)-f(x)}{t-x}=\frac{t^n-x^n}{t-x}=\frac{(t-x)\left(t^{n-1}+t^{n-2}x+\ldots+tx^{n-2}+x^{n-1}\right)}{t-x}$$

$$=t^{n-1}+t^{n-2}x+\ldots+tx^{n-2}+x^{n-1} \text{ if } x\neq p.$$

Taking limits as $t\to x$, we obtain

$$\lim_{t\to x}\frac{f(t)-f(x)}{t-x}=x^{n-1}+x^{n-2}x+\ldots+xx^{n-2}+x^{n-1}.$$

Therefore, $f'(x)=nx^{n-1}$, $x\in\mathbb{R}$, where $n\in\mathbb{N}$.

Note: If $n=0$, i.e., $f(x)=1$ or $f(x)=c\in\mathbb{R}$ for all $x\in\mathbb{R}$, then $f'(x)=0$ for all $x\in\mathbb{R}$.

4.1E2 Example: Let $f:\mathbb{R}\to\mathbb{R}$ be defined by

$$f(x)=\begin{cases} x, & x\geq 0 \\ -x, & x<0. \end{cases}$$

Then

$$f_+'(0)=\lim_{x\to 0+}\frac{f(x)-f(0)}{x-0}=\lim_{x\to 0+}\frac{|x|}{x}=\lim_{x\to 0+}\frac{x}{x}=1$$

and

$$f_-'(0)=\lim_{x\to 0-}\frac{f(x)-f(0)}{x-0}=\lim_{x\to 0-}\frac{|x|}{x}=\lim_{x\to 0-}\frac{-x}{x}=-1.$$

Therefore, f is not differentiable at 0. However, f is differentiable at all other points of $\mathbb{R}$.

Note: From the above example we see that a function that is continuous at a point may not be differentiable at that point. However, a function that is differentiable is continuous as we see from the next theorem.

4.1T1 Theorem:

Suppose $f: I \to \mathbb{R}$ is differentiable at $p \in I,$ where *I* is an interval in $\mathbb{R}$. Then *f* is continuous at *p*.

Proof: Suppose $x \neq p$. Then

$$\lim_{x\to p}\left[f(x)-f(p)\right]=\lim_{x\to p}\left[\frac{f(x)-f(p)}{x-p}\right]\lim_{x\to p}(x-p)=f'(p)\cdot 0=0.$$

Hence result.

4.1T2 Theorem:

Suppose $f, g: I \to \mathbb{R}$ are differentiable at a point $p \in \mathbb{R},$ where *I* is an interval in $\mathbb{R}$. Then $f+g,\ fg,$ and f/g $\left(\text{if } g(p)\neq 0\right)$ are differentiable at *p* and

(a) $(f+g)'(p)=f'(p)+g'(p)$

(b) $(fg)'(p)=f'(p)g(p)+f(p)g'(p)$

(c) $\left(\dfrac{f}{g}\right)'(p)=\dfrac{f'(p)g(p)-f(p)g'(p)}{\left[g(p)\right]^2}$ if $g(p)\neq 0.$

Proof:

(a) Follows immediately using (3.3T1) and (4.1D1).

(b) For $x \neq p,$

$$\frac{f(x)g(x)-f(p)g(p)}{x-p}=\frac{f(x)-f(p)}{x-p}g(x)+f(p)\frac{g(x)-g(p)}{x-p}.$$

Since *g* is continuous at *p* (4.1T1), we can use (3.3T1) and (4.1D1) to obtain

$$(fg)'(p)=f'(p)g(p)+f(p)g'(p).$$

(c) Let $h=f/g,\ g(p)\neq 0,$ and $x \neq p.$ Then

$$\frac{h(x)-h(p)}{x-p}=\frac{1}{g(x)g(p)}\left[\frac{f(x)-f(p)}{x-p}g(p)-f(p)\frac{g(x)-g(p)}{x-p}\right].$$

Since *g* is continuous at *p* (4.1T1), we can use (3.3T1) and (4.1D1) to obtain

$$\left(\frac{f}{g}\right)'(p)=\frac{f'(p)g(p)-f(p)g'(p)}{\left[g(p)\right]^2}.$$

Note: If we take $f(x) \equiv 1$ in (c), then we obtain: $\left(\frac{1}{g}\right)'(p) = \frac{-g'(p)}{[g(p)]^2}$.

4.1L1 Lemma:

Let $f: I \to \mathbb{R}$ where I is an interval in $\mathbb{R}$. Then f is differentiable at a point p in I if and only if there exists a function $\eta: I \to \mathbb{R}$ such that $\lim_{x \to p} \eta(x) = 0 = \eta(p)$ and

$$\frac{f(x) - f(p)}{x - p} = L + \eta(x) \text{ for } x \neq p,$$

where $L \in \mathbb{R}$.

Proof: Suppose f is differentiable at p. Define $\eta: I \to \mathbb{R}$ by

$$\eta(x) = \begin{cases} \frac{f(x) - f(p)}{x - p} - f'(p), & x \in (I - \{p\}) \\ 0, & x = p. \end{cases}$$

Then $\lim_{x \to p} \eta(x) = 0$ and

$$\frac{f(x) - f(p)}{x - p} = f'(p) + \eta(x) \text{ for } x \neq p.$$

Suppose there exists a function $\eta: I \to \mathbb{R}$ such that $\lim_{x \to p} \eta(x) = 0 = \eta(p)$ and

$$\frac{f(x) - f(p)}{x - p} = L + \eta(x) \text{ for } x \neq p,$$

where $L \in \mathbb{R}$. Then

$$\lim_{x \to p} \left[\frac{f(x) - f(p)}{x - p} - L \right] = \lim_{x \to p} \eta(x) = 0.$$

Therefore f is differentiable at p.

4.1T3 Theorem: Chain Rule

Suppose $f: I \to \mathbb{R}$, $g: J \to \mathbb{R}$ where I and J are intervals in $\mathbb{R}$ such that $Ran(f) \subseteq J$. If f is differentiable at $p \in I$ and g is differentiable at $f(p)$, then the function $g \circ f$ is differentiable at p and

$$h'(p) = g'(f(p)) f'(p).$$

Proof: Let $q = f(p)$. Then by (4.1L1), there exist functions $\eta: I \to \mathbb{R}$ and $\mu: J \to \mathbb{R}$ such that $\lim_{x \to p} \eta(x) = 0 = \eta(p)$, $\lim_{y \to f(p)} \mu(y) = 0 = \mu(f(p))$,

$$f(x) - f(p) = (x - p)[f'(p) + \eta(x)],$$

and

$$g(y)-g(q)=(y-q)\left[g'(q)+\mu(y)\right].$$

Let $y=f(x)$ and $q=f(p)$. Using the above, we obtain

$$\begin{aligned}h(x)-h(p)&=g(f(x))-g(f(p))\\&=g(y)-g(f(p))\\&=\left[y-f(p)\right]\left[g'(f(p))+\mu(y)\right]\\&=\left[f(x)-f(p)\right]\left[g'(f(p))+\mu(y)\right]\\&=(x-p)\left[f'(p)+\eta(x)\right]\left[g'(f(p))+\mu(y)\right].\end{aligned}$$

If $x\neq p$, then

$$\frac{h(x)-h(p)}{x-p}=\left[f'(p)+\eta(x)\right]\left[g'(f(p))+\mu(y)\right].$$

Since *f* is continuous at *p*,

$$\lim_{x\to p}\mu(f(x))=\mu\left(\lim_{x\to p}f(x)\right)=\mu(f(p))=0.$$

Therefore,

$$\lim_{x\to p}\frac{h(x)-h(p)}{x-p}=f'(p)g'(f(p)).$$

4.1E3 Example: Let $f:\mathbb{R}\to\mathbb{R}$ be defined by

$$f(x)=\begin{cases}x^3+1, & x\leq 1\\ ax+b, & x>1.\end{cases}$$

(a) For what values of *a* and *b* is *f* continuous at 1?

(b) For what values of *a* and *b* is *f* differentiable at 1?

(a) $f(1)=2,\ \lim_{x\to 1-}f(x)=1^3+1=2$ and $\lim_{x\to 1+}f(x)=a+b$. For continuity at 1, we need $a+b=2,\ a,b\in\mathbb{R}$.

(b) For differentiability, we need the above condition in addition to $f_-'(1)=f_+'(1)$, i.e.,

$$\begin{aligned}f_-'(1)=\lim_{x\to 1-}\frac{(x^3+1)-2}{x-1}&=\lim_{x\to 1-}(x^2+x+1)\\&=3=a\\&=\lim_{x\to 1+}\frac{(ax+b)-(a+b)}{x-1}=f_+'(1).\end{aligned}$$

Therefore, $a=3$ and $b=-1$.

4.1A1 Assignment: Let $f:\mathbb{R}\to\mathbb{R}$ be defined by

$$f(x)=\begin{cases} ax+b, & x<-1 \\ x^3+1, & -1\le x\le 1 \\ cx+d, & x>1. \end{cases}$$

Determine *a, b, c,* and *d* such that *f* is differentiable on $\mathbb{R}$.

4.1A2 Assignment:

Let $f:\mathbb{R}\to\mathbb{R}$ be defined by $f(x)=\sqrt{x},\ x\ge 0$. Show that for all $x>0$,

$$f'(x)=\frac{1}{2\sqrt{x}}.$$

4.1D3 Definition: Derivatives of Higher Order

Suppose $f:I\to\mathbb{R}$ is differentiable on *I*, where *I* is an interval. If f' is differentiable at a point *p* in *I*, then its derivative at *p* is called the **second derivative of *f* at *p*** and is denoted $f''(p)$ or $f^{(2)}(p)$. The **second derivative of *f*** is the function $f'':E\to\mathbb{R}$, given by

$$f''(x)=\lim_{t\to x}\frac{f'(t)-f'(x)}{t-x},\ \text{for all } x\in E,$$

where $E=\{x\in I \mid f''(x) \text{ exists}\}$.

We can extend this definition to define the ***n*****th derivative of *f* at *p*** that we will denote $f^{(n)}(p)$ when $n\ge 3$. The ***n*****th derivative of *f*** is the function $f^{(n)}:E\to\mathbb{R}$, given by

$$f^{(n)}(x)=\lim_{t\to x}\frac{f^{(n-1)}(t)-f^{(n-1)}(x)}{t-x},\ \text{for all } x\in E,$$

where $E=\{x\in I \mid f''(x) \text{ exists}\}$.

Notation: If we use $y=f(x)$ to denote the function *f*, then its first derivative is denoted by $\frac{dy}{dx}$ and its second derivative is denoted by $\frac{d^2y}{dx^2}$. The nth derivative of *f* is denoted by $\frac{d^ny}{dx^n}$. The nth derivative of *f* at a point *p* is denoted by $\left.\frac{d^ny}{dx^n}\right|_{x=p}$.

4.1E4 Example: Suppose $f:\mathbb{R}\to\mathbb{R}$ is defined by $f(x)=x^n,\ x\in\mathbb{R}$, where $n\in\mathbb{N}$. Use mathematical induction to show that the kth derivative of *f* exists for all $k\in\mathbb{N}$ and is given by:

$$f^{(k)}(x)=\begin{cases} n(n-1)\ldots(n-k+1)x^{n-k}, & \text{if } k\le n \\ 0, & \text{if } k>n. \end{cases}$$

We have shown that $f'(x) = nx^{n-1}$, $x \in \mathbb{R}$. Suppose

$$f^{(m)}(x) = n(n-1)...(n-m+1)x^{n-m}, \quad x \in \mathbb{R}.$$

Then

$$f^{(m+1)}(x) = n(n-1)...(n-m+1)(n-(m+1)-1)x^{n-(m+1)}, \quad x \in \mathbb{R}.$$

Therefore,

$$f^{(k)}(x) = n(n-1)...(n-k+1)x^{n-k} \text{ for all } k = 1,2,....n$$

Since

$$f^{(n)}(x) = n(n-1)...(n-n+1)x^{n-n} = n!,$$

it follows that

$$f^{(k)}(x) = 0 \text{ for all } x \in \mathbb{R} \text{ when } k > n.$$

4.1A3 Assignment:

(1) Suppose $L:(0,\infty) \to \mathbb{R}$ satisfies $L'(x) = 1/x$ for all $x \in (0,\infty)$. Find the derivative of each of the following functions:

(a) $f(x) = L(3x-2), \quad x > \dfrac{2}{3}.$

(b) $g(x) = L(x^2), \quad x \neq 0.$

(c) $h(x) = [L(x)]^4, \quad x > 0.$

(2) If f is differentiable at p, then show that

$$\lim_{h \to 0} \frac{f(p+h) - f(p-h)}{2h} = f'(p).$$

Is the converse true? Justify your answer!

4.2 Mean Value Theorems

We will discuss various versions of the mean value theorem in this section. These include Rolle's theorem and the extended mean value theorem.

4.2D1 Definition:

Let $f: E \to \mathbb{R}$ where $E \subseteq \mathbb{R}$. Then f is said to have a **relative (local) maximum** at a point p of E if there exists $\delta > 0$ such that

$$f(x) \leq f(p) \text{ for all } x \in E \cap N_\delta(p).$$

The function f is said to have a **relative (local) minimum** at a point p of E if there exists $\delta > 0$ such that

$$f(x) \geq f(p) \text{ for all } x \in E \cap N_\delta(p).$$

The function *f* has an **absolute maximum at *p*** if

$$f(x) \leq f(p) \text{ for all } x \in E$$

and *f* has an **absolute minimum at *p*** if

$$f(x) \geq f(p) \text{ for all } x \in E.$$

The function *f* has a **relative extremum** at a point *p* of *E* if it has a relative maximum or a relative minimum at *p*. An **absolute extremum** is an absolute maximum or an absolute minimum.

Note: Every continuous real valued function defined on a compact subset *K* of $\mathbb{R}$ has an absolute maximum and an absolute minimum on *K*.

4.2T1 Theorem:

Suppose $f : I \to \mathbb{R}$ is differentiable at a point $p \in \mathbb{R}$, where *I* is an interval in $\mathbb{R}$ and *f* has a relative extremum at an interior point *p* of *I*. If *f* is differentiable at *p*, then $f'(p) = 0$.

Proof: If *f* is differentiable at *p* then both $f'_-(p)$ and $f'_+(p)$ exist and are equal. Without loss of generality we may assume that *f* has a relative maximum at *p*. Then there exists $\delta > 0$ such that $N_\delta(p) \subseteq I$ and

$$f(x) \leq f(p) \text{ for all } x \in N_\delta(p).$$

Thus if $p < x < p + \delta$, then

$$\frac{f(x) - f(p)}{x - p} \leq 0,$$

and if $p - \delta < x < p$, then

$$\frac{f(x) - f(p)}{x - p} \geq 0.$$

Therefore, $f'_+(p) \leq 0$ and $f'_-(p) \geq 0$. Hence, $f'(p) = 0$ since $f'_-(p) = f'_+(p) = f(p)$.

4.2T1C Corollary: Let $f : [a,b] \to \mathbb{R}$ be continuous where $a \quad b$ If *f* has a relative extremum at $p \in (a,b)$, then either $f'(p) = 0$ or $f'(p)$ does not exist.

Proof: Follows immediately.

4.2T2 Theorem: Rolle's Theorem

Suppose $a < b$, $f : [a,b] \to \mathbb{R}$ is continuous, differentiable on (a,b), and $f(a) = f(b)$. Then there exists $c \in (a,b)$ such that $f'(c) = 0$.

Proof: If f is constant, then $f'(x)=0$ for all $x\in(a,b)$. Suppose f is not constant. Since f is continuous, it has its absolute extrema on $[a,b]$. If $f(t)>f(a)$ for some $t\in(a,b)$, then f has its absolute maximum at some $c\in(a,b)$. This absolute maximum is also a relative maximum. Thus, by the previous theorem, $f'(c)=0$. Similarly, if $f(t)<f(a)$ for some $t\in(a,b)$, then it follows that f has an absolute minimum at some $c\in(a,b)$ and $f'(c)=0$.

4.2E1 Example: Show that the equation $x^3-3x+a=0$ has at most one root in $[-1,1]$.

Let $f:[-1,1]\to\mathbb{R}$ be defined by: $f(x)=x^3-3x+a$. Suppose $f(a)=f(b)$ for some $a,b\in[-1,1]$ with $a<b$. Then since f is continuous and differentiable on $[a,b]$, there exists $c\in(a,b)$ such that $f'(c)=0$. But $f'(x)=3x^2-3\neq 0$ for any $x\in(-1,1)$. Since $a,b\in[-1,1]$, c must be in $(-1,1)$. This is a contradiction. Hence result.

4.2T3 Theorem: Generalized Mean Value Theorem

Suppose $a<b$, $f,g:[a,b]\to\mathbb{R}$ are continuous, and differentiable on (a,b). Then there exists $c\in(a,b)$ such that

$$[g(b)-g(a)]f'(c)=[f(b)-f(a)]g'(c).$$

Proof: Define $h:[a,b]\to\mathbb{R}$ by

$$h(x)=[g(b)-g(a)]f(x)-[f(b)-f(a)]g(x),\ x\in[a,b].$$

Then h is continuous, differentiable on (a,b), and

$$h(a)=h(b)=f(a)g(b)-g(a)f(b).$$

Therefore, by Rolle's theorem, there exists $c\in(a,b)$ such that $h'(c)=0$. But for all $x\in(a,b)$,

$$h'(x)=[g(b)-g(a)]f'(x)-[f(b)-f(a)]g'(x).$$

Hence result.

4.2T4 Theorem: Mean Value Theorem (MVT)

Suppose $a<b$, $f:[a,b]\to\mathbb{R}$ is continuous, and differentiable on (a,b). Then there exists $c\in(a,b)$ such that

$$f'(c)=\frac{f(b)-f(a)}{b-a}.$$

Proof: Define $g:[a,b]\to\mathbb{R}$ by $g(x)=x,\ x\in[a,b]$. Now apply the previous theorem to obtain $(b-a)f'(c)=[f(b)-f(a)]\cdot 1$. Hence result.

4.2E2 Example: Suppose $f,g:[0,\infty)\to\mathbb{R}$ are continuous and differentiable on $(0,\infty)$. If $f(0)=g(0)$ and $f'(x)\le g'(x)$ for all $x\in(0,\infty)$, prove that $f(x)\le g(x)$ for all $x\in[0,\infty)$.

Define $h:[0,\infty)\to\mathbb{R}$ by $h(x)=g(x)-f(x),\ x\in[0,\infty)$.

Let $x\in(0,\infty)$. Then h is continuous on $[0,x)$, and differentiable on $(0,x)$. Therefore, by the MVT, there exists $c\in(0,x)$ such that $h'(c)=\dfrac{h(x)-h(0)}{x-0}$. But $h'(c)=g'(c)-f'(c)\ge 0$, $x>0$, and $h(0)=g(0)-f(0)=0$. This implies that $h(x)\ge 0$. Hence result.

4.2E3 Example: Suppose $f:I\to\mathbb{R}$ is differentiable on *I*, where *I* is an interval. Prove that f' is bounded on *I* if and only if there exists $M>0$ such that

$$|f(x)-f(y)|\le M|x-y| \text{ for all } x,y\in I.$$

Suppose f' is bounded on *I*. Then there exists $M>0$ such that $|f'(x)|\le M$ for all $x\in I$. Suppose $x,y\in I$. Without loss of generality, we may assume that $y<x$. Then since *f* is differentiable on $[y,x]$, by the MVT, there exists $c\in I$ such that

$$f(x)-f(y)=f'(c)(x-y).$$

Therefore,

$$|f(x)-f(y)|=|f'(c)||x-y|\le M|x-y|.$$

Conversely, suppose there exists $M>0$ such that

$$|f(x)-f(y)|\le M|x-y| \text{ for all } x,y\in I.$$

Let $x,y\in I$ with $y\ne x$. Then

$$|f'(x)|=\left|\lim_{y\to x}\frac{f(x)-f(y)}{x-y}\right|=\lim_{y\to x}\left|\frac{f(x)-f(y)}{x-y}\right|\le M \text{ for all } x\in I.$$

Note that if $\lim_{x\to p}f(x)$ exists, then so does $\lim_{x\to p}|f(x)|$ and $\lim_{x\to p}|f(x)|=\left|\lim_{x\to p}f(x)\right|$.

4.2T5 Theorem: Suppose $f:I\to\mathbb{R}$ is differentiable on *I*, where *I* is an interval.

(a) If $f'(x)\ge 0$ for all $x\in I$, then *f* is increasing on *I*.

(b) If $f'(x) > 0$ for all $x \in I^{\circ}$, then *f* is strictly increasing on *I*.

(c) If $f'(x) \le 0$ for all $x \in I$, then *f* is decreasing on *I*.

(d) If $f'(x) < 0$ for all $x \in I^{\circ}$, then *f* is strictly decreasing on *I*.

(e) If $f'(x) = 0$ for all $x \in I^{\circ}$, then *f* is constant on *I*.

Proof: Suppose $x_1, x_2 \in I$ with $x_1 < x_2$. We can now apply the Mean Value Theorem to *f* on the interval $[x_1, x_2]$, since the hypotheses are satisfied. Thus,

$$f(x_2) - f(x_1) = f'(c)(x_1 - x_2) \text{ for some } c \in (x_1, x_2).$$

(a) $f'(c) \ge 0 \Rightarrow f(x_2) - f(x_1) \ge 0 \Rightarrow f(x_2) \ge f(x_1)$. Hence, *f* is increasing on *I*.

(b) $f'(c) > 0 \Rightarrow f(x_2) - f(x_1) > 0 \Rightarrow f(x_2) > f(x_1)$. Hence, *f* is strictly increasing on *I*.

(c) $f'(c) \le 0 \Rightarrow f(x_2) - f(x_1) \le 0 \Rightarrow f(x_2) \le f(x_1)$. Hence, *f* is decreasing on *I*.

(d) $f'(c) < 0 \Rightarrow f(x_2) - f(x_1) < 0 \Rightarrow f(x_2) < f(x_1)$. Hence, *f* is strictly decreasing on *I*.

(e) $f'(c) = 0 \Rightarrow f(x_2) - f(x_1) = 0 \Rightarrow f(x_2) = f(x_1)$. Hence, *f* is constant on *I*.

Note: It is easily seen that if $f'(c) > 0$ for some $c \in I$, then there exists $\delta > 0$ such that for all $x \in (c - \delta, c)$, $f(x) < f(c)$ and for all $x \in (c, c + \delta)$, $f(x) > f(c)$. This does not mean that *f* is strictly increasing on *I*.

4.2 T6 Theorem: First Derivative Test for Relative Extrema

Suppose $a < b$, $c \in (a,b)$, $f : (a,b) \to \mathbb{R}$ is continuous, and *f* is differentiable on (a,c) and (c,b). The derivative of *f* need not exist at *c*.

(a) If $f'(x) > 0$ for all $x \in (a,c)$ and $f'(x) < 0$ for all $x \in (c,b)$, then *f* has a relative maximum at *c*.

(b) If $f'(x) < 0$ for all $x \in (a,c)$ and $f'(x) > 0$ for all $x \in (c,b)$, then *f* has a relative minimum at *c*.

Proof:

(a) By (4.2T5), *f* is strictly increasing on $(a,c]$ and strictly decreasing on $[c,b)$. Therefore, for all $x \in (a,b)$, $f(x) \le f(c)$. Hence, *f* has a relative maximum at *c*.

(b) By (4.2T5), f is strictly decreasing on $(a,c]$ and strictly increasing on $[c,b)$. Therefore, for all $x \in (a,b)$, $f(x) \geq f(c)$. Hence, f has a relative minimum at c.

4.2T7 Theorem:

Suppose $f:[a,b] \to \mathbb{R}$ is continuous, and differentiable on (a,b), where $a < b$. Then the one-sided derivative $f_+'(a)$ exists if $\lim\limits_{x \to a+} f'(x)$ exists. In this case, $f_+'(a) = \lim\limits_{x \to a+} f'(x)$. Furthermore, if $\lim\limits_{x \to b-} f'(x)$ exists, then $f_-'(b)$ exists and equals $\lim\limits_{x \to b-} f'(x)$.

Proof: Suppose $\lim\limits_{x \to a+} f'(x) = L \in \mathbb{R}$. If $\varepsilon > 0$ is given, then there exists $\delta > 0$ such that $\delta < b - a$ and $|f'(x) - L| < \varepsilon$ for all $x \in (a, a+\delta)$. Suppose $t \in (a, a+\delta)$. Since f is continuous on $[a,t]$ and differentiable on (a,t), by the MVT, there exists $c \in (a,t)$ such that

$$f(t) - f(a) = (t-a) f'(c).$$

Thus

$$\left| \frac{f(t) \quad f(a)}{t \quad a} - L \right| = |f'(c) - L| < \quad \text{for all } t \in (a, a+\delta).$$

This establishes $f_+'(a) = \lim\limits_{x \to a+} f'(x)$. Similarly, we can show that $f_-'(b) = \lim\limits_{x \to b-} f'(x)$.

4.2E4 Example: Consider the function $f:\mathbb{R} \to \mathbb{R}$ defined by

$$f(x) = \begin{cases} x^2 + 1, & x < 1 \\ 2x, & x \geq 1. \end{cases}$$

f is clearly continuous. It is quite clear that f is differentiable on $\mathbb{R} \setminus \{1\}$. Furthermore,

$$f'(x) = \begin{cases} 2x, & x < 1 \\ 2, & x > 1. \end{cases}$$

Therefore,

$$\lim_{x \to 1-} f'(x) = 2 = \lim_{x \to 1+} f'(x).$$

It follows that

$$f_-'(1) = f_+'(1) = 2.$$

Hence, f is also differentiable at 1 and $f'(1) = 2$.

4.2A1 Assignment:

1. Use the Mean Value Theorem to establish the following inequalities:

   (a) $\sqrt{1+x} \leq 1 + x/2$ for all $x > -1$.

(b) $(1+x)^n \geq 1+nx$ for all $x > -1$ where $n \in \mathbb{N}$.

2. Suppose f is continuous on $[0,\infty)$, $f(0)=0$, f is differentiable on $(0,\infty)$, and f' is increasing on $(0,\infty)$. Define $g:(0,\infty \to \mathbb{R})$ by $g(x)=\dfrac{f(x)}{x}$ for all $x>0$. Prove that g is increasing on $(0,\infty)$.

3. Suppose $f:\mathbb{R}\to\mathbb{R}$ is continuous, and differentiable on $\mathbb{R}\setminus\{0\}$. If $\lim_{x\to 0} f'(x)=L$, then prove that f is differentiable at 0 and $f'(0)=L$.

4. Show that the equation $x^4+4x^2-a=0$ can have at most two distinct real roots.

5. Suppose $f:\mathbb{R}\to\mathbb{R}$ satisfies $|f(x)-f(y)| \leq M|x-y|^{\alpha}$ for some $\alpha>1$ and $x,y\in\mathbb{R}$. Prove that f is a constant function.

4.2T8 Theorem: Intermediate Value Theorem for Derivatives

Suppose $f:I\to\mathbb{R}$ is differentiable on I, where I is an interval. Let $a,b\in I$ with $a<b$. If λ is a real number between $f'(a)$ and $f'(b)$, then there exists $c\in(a,b)$ such that $f'(c)=\lambda$.

Proof: Define $g:I\to\mathbb{R}$ by

$$g(x)=f(x)-\lambda x \text{ for all } x\in I.$$

Then g is differentiable on I and

$$g'(x)=f'(x)-\lambda \text{ for all } x\in I.$$

Suppose $f'(a)<\lambda<f'(b)$. Then

$$g'(a)<0 \text{ and } g'(b)>0.$$

Thus

$$g(x_1)<g(a) \text{ and } g(x_2)<g(b)$$

for some $x_1\in(a,b)$ and $x_2\in(a,b)$. (See the note after (4.2T5).) It follows that

$\min_{[a,b]} g = g(c)$ for some $c\in(a,b)$ and $g'(c)=0$. Hence $f'(c)=\lambda$.

4.2T8C Corollary: Suppose $f: I \to \mathbb{R}$ is differentiable on *I*, where *I* is an interval and let $a, b \in I$ with $a < b$. If $f'(a) = 0 = f'(b)$ and $f'(x) \neq 0$ for all $x \in (a,b)$, then f' has the same sign on (a,b).

Proof: Suppose $f'(c) > 0$ for some $c \in (a,b)$. If this is not the case, then $f'(x) < 0$ for all $x \in (a,b)$ and the result is true. If $f'(x) < 0$ for some $x \in (a,b)$, then by the above theorem, there exists $\gamma \in (a,b)$ such that $f'(\gamma) = 0$. This is a contradiction. Therefore, $f'(x) > 0$ for all $x \in (a,b)$.

4.2T9 Theorem: Inverse Function Theorem

Suppose $f: I \to \mathbb{R}$ is differentiable on *I*, where *I* is an interval and $f'(x) \neq 0$ for all $x \in I$. Then *f* is one-to-one on I, its inverse f^{-1} is one-to-one and differentiable on $J = f(I)$, and

$$\left(f^{-1}\right)'\left(f(x)\right) = \frac{1}{f'(x)} \text{ for all } x \in I.$$

Proof: Since $f'(x) \neq 0$ for all $x \in I$, by (4.2T8C), f' is either positive or negative on *I*. Without loss of generality, we may assume that it is positive on *I*. Then by (4.2T5(b)), *f* is strictly increasing on *I*. Let $y \in J$ and let $\{y_n\}$ be any sequence in *J* with $y_n \neq y$ for all $n \in N$ and $\lim_{n\to\infty} y_n = y$. Then there exists a sequence $\{x_n\}$ in *I* such that $y_n = f(x_n)$ for all $n \in \mathbb{N}$. Since f^{-1} is continuous,

$$\lim_{n\to\infty} x_n = \lim_{n\to\infty} f^{-1}(y_n) = f^{-1}(y) = x.$$

Hence,

$$\lim_{n\to\infty} \frac{f^{-1}(y_n) - f^{-1}(y)}{y_n - y} = \lim_{n\to\infty} \frac{x_n - x}{f(x_n) - f(x)} = \frac{1}{f'(x)}.$$

Since this holds for any sequence $\{y_n\}$ in *J* with $y_n \neq y$ for all $n \in N$ and $\lim_{n\to\infty} y_n = y$, it follows that

$$\left(f^{-1}\right)'\left(f(x)\right) = \left(f^{-1}\right)'(y) = \lim_{t\to y} \frac{f^{-1}(t) - f^{-1}(y)}{t - y} = \frac{1}{f'(x)}.$$

4.2E5 Example: Define $f:(0,\infty) \to \mathbb{R}$ by $f(x) = x^r$, $x \in (0,\infty)$ where $r = \frac{m}{n} \in \mathbb{Q}$ in reduced form. Show that $f'(x) = rx^{r-1}$, for $x \in (0,\infty)$. It is assumed that $x^{r-1} \equiv 0$ if $r = 1$.

Define $g:(0,\infty) \to \mathbb{R}$ by

$$g(x) = x^n, \; x \in (0,\infty)$$

where $n \in \mathbb{N}$. Then $g^{-1}:(0,\infty) \to \mathbb{R}$ is given by

$$g^{-1}(x) = x^{1/n},\ x \in (0,\infty)$$

where $n \in \mathbb{N}$. But

$$g'(x) = nx^{n-1},\ \ x \in (0,\infty).$$

Therefore,

$$\left(g^{-1}\right)'(x) = \frac{1}{g'\left(g^{-1}(x)\right)}$$

$$= \frac{1}{n\left(x^{1/n}\right)^{n-1}} = \frac{1}{n}x^{1/n-1}.$$

Now define $h:(0,\infty) \to \mathbb{R}$ by

$$h(x) = x^m,\ \ x \in (0,\infty)$$

where $m \in \mathbb{N}$. Then $f = g^{-1} \circ h$ and

$$f'(x) = \left(g^{-1}\right)'\left(h(x)\right)h'(x)$$

$$= \frac{1}{n}\left(x^m\right)^{1/n-1} mx^{m-1} = \frac{m}{n}x^{m/n-1}$$

$$= rx^{r-1},\ x \in (0,\infty).$$

4.2E6 Example: Suppose $L:(0,\infty) \to \mathbb{R}$ satisfies $L'(x) = 1/x,\ x \in (0,\infty)$. Prove that L has an inverse defined on $L\left((0,\infty)\right)$ and $\left(L^{-1}\right)'(x) = L^{-1}(x)$ for all $x \in L\left((0,\infty)\right)$.

Since $L:(0,\infty) \to \mathbb{R}$ satisfies $L'(x) = 1/x,\ x \in (0,\infty)$, L is strictly increasing, one-to-one, and differentiable on $(0,\infty)$. Hence, L^{-1} exists and is differentiable on $L\left((0,\infty)\right)$. Let $x \in (0,\infty)$. If $y = L(x)$, then

$$\left(L^{-1}\right)'(y) = 1/L'(x) = x = L^{-1}(y) \text{ for all } y \in L\left((0,\infty)\right).$$

4.2T10 Thoerem: L'Hospital's Rule

Suppose $f, g: I \to \mathbb{R}$ are differentiable on I with $g(x) \neq 0$ for all $x \in I$ and

$$\lim_{x \to a}\frac{f'(x)}{g'(x)} = L \in \mathbb{R} \cup \{-\infty, \infty\},$$

where $a \in I$. If $\lim_{x \to a} f(x) = \lim_{x \to a} g(x) = 0$, then

$$\lim_{x \to a}\frac{f(x)}{g(x)} = L.$$

Proof: Suppose $a \in I$, $(a, a+\delta) \subseteq I$ for some $\delta > 0$, and $L < \infty$. Choose $q, r \in \mathbb{R}$ such that $L \le r < q$. Then there exists $c \in (a, a+\delta)$ such that

$$x \in (a,c) \Rightarrow \frac{f'(x)}{g'(x)} < r.$$

If $x, y \in (a,c)$ with $x < y$, then there exists $t \in (x,y)$ such that

$$\frac{f(y)-f(x)}{g(y)-g(x)} = \frac{f'(t)}{g'(t)} < r.$$

If $\lim_{x \to a} f(x) = \lim_{x \to a} g(x) = 0$, then let $x \to a$ in the previous identity to obtain

$$\frac{f(y)}{g(y)} \le r < q \text{ for all } y \in (a,c).$$

Similarly, for $a \in I$, $(a, a+\delta) \subseteq I$ for some $\delta > 0$, and $L > -\infty$, there exists $p \in \mathbb{R}$ such that

$$p < r \le L \text{ and } p < r \le \frac{f(y)}{g(y)} \text{ for all } y \in (a,c).$$

Thus $\lim_{x \to a+} \frac{f(x)}{g(x)} = L$. In a similar manner, we can show that $\lim_{x \to a-} \frac{f(x)}{g(x)} = L$.

Note: The above proof covers the cases where $a = \pm\infty$ as well as $L = \pm\infty$.

Note: The above result holds even if $\lim_{x \to a} f(x) = \lim_{x \to a} g(x) = 0$ is replaced by $\lim_{x \to a} g(x) = \pm\infty$.

4.2T11 Theorem: Taylor's Theorem

Suppose $f, f', \ldots, f^{(n)} : [a,b] \to \mathbb{R}$ are continuous and $f^{(n)}$ is differentiable on (a,b), where $n \in \mathbb{N}$. Let $\alpha, \beta \in [a,b]$ with $\alpha \ne \beta$ and define

$$P(t) = \sum_{k=0}^{n} \frac{f^{(k)}(\alpha)}{k!}(t-\alpha)^k \text{ for all } t \in [a,b].$$

Then there exists c beteween α and β such that

$$f(\beta) = P(\beta) + \frac{f^{(n+1)}(c)}{(n+1)!}(\beta-\alpha)^{n+1}.$$

Proof: Define M by

$$f(\beta) = P(\beta) + M(\beta-\alpha)^{n+1}$$

and let

$$g(t) = f(t) - P(t) - M(t-\alpha)^{n+1} \text{ for } t \in [a,b].$$

Then

$$g^{(n+1)}(t) = f^{(n+1)}(t) - (n+1)!M \text{ for } t \in (a,b).$$

Since

$$P^{(k)}(\alpha) = f^{(k)}(\alpha) \text{ for } k = 1,2,...,n,$$

we have

$$g(\alpha) = g'(\alpha) = ... = g^{(n)}(\alpha) = 0.$$

Furthermore, $g(\beta) = 0,$ so that $g'(x_1) = 0$ for some x_1 between α and β by the MVT. Since $g'(\alpha) = 0,$ we obtain $g''(x_2) = 0$ for some x_2 between α and x_1 by the MVT. After $(n+1)$ steps we arrive at the conclusion $g^{(n+1)}(c) = 0$ for some c between α and x_n by the MVT. Therefore,

$$f^{(n+1)}(c) = (n+1)!M$$

and

$$f(\beta) = P(\beta) + \frac{f^{(n+1)}(c)}{(n+1)!}(\beta - \alpha)^{n+1}.$$

4.2E7 Example: Suppose $f:(a,\infty) \to \mathbb{R}$ is twice differentiable on its domain, where $a \in \mathbb{R}$. Define $M_0 = \sup_{x \in (a,\infty)} |f(x)|$, $M_1 = \sup_{x \in (a,\infty)} |f'(x)|$, $M_2 = \sup_{x \in (a,\infty)} |f''(x)|$. Prove that $M_1^2 \le 4M_1M_2$.

Let $x \in (a,\infty)$ and $h > 0$. Apply Taylor's theorem with $\alpha = x$ and $\beta = x + 2h$ to obtain

$$P(t) = \sum_{k=0}^{1} \frac{f^{(k)}(x)}{k!}(t-x)^k = f(x) + 2hf'(x)$$

and

$$f(x+2h) = f(x) + 2hf'(x) + \frac{(2h)^2}{2!} f''(c)$$

for some $c \in (x, x+2h)$. This gives us

$$\begin{aligned} f'(x) &= \frac{[f(x+2h) - f(x)]}{2h} - hf''(c) \\ &\le \frac{2M_0}{2h} + hM_2. \end{aligned}$$

Since this inequality holds for all $x \in (a,\infty)$ and $h > 0,$

$$M_1 \le \frac{M_0}{h} + hM_2.$$

$$\Rightarrow M_2h^2 - M_1h + M_0 \ge 0 \text{ for all } h > 0.$$

Therefore, the discriminant of the quadratic in the above inequality must be nonpositive, i.e.,

$$M_1^2 - 4M_0M_2 \le 0.$$

Hence,

$$M_1^2 \le 4M_0M_2.$$

4.2A2 Assignment: Suppose f is twice differentiable on $(0,\infty)$ and f, f', and f'' are bounded on $(0,\infty)$. If $\lim_{x\to\infty} f(x) = 0$, then use the above example to show that $\lim_{x\to\infty} f'(x) = 0$.

4.2 E8 Example: Suppose $f:[-1,1] \to \mathbb{R}$ is three times differentiable on its domain and

$$f(-1) = f(0) = 0,\ f(1) = 1, \text{ and } f'(0) = 0.$$

Prove that $f'''(c)$ for some $c \in (-1,1)$.

Since $f(-1) = f(0) = 0,\ f(1) = 1, \text{ and } f'(0) = 0$, it seems reasonable to use Taylor's theorem with $\alpha = 0$ and $\beta = \pm 1$. This gives us:

$$f(-1) = f(0) + \frac{f'(0)}{1!}(-1-0)^1 + \frac{f''(0)}{2!}(-1-0)^2 + \frac{f'''(c_1)}{3!}(-1-0)^3$$

and

$$f(1) = f(0) + \frac{f'(0)}{1!}(1-0)^1 + \frac{f''(0)}{2!}(1-0)^2 + \frac{f'''(c_2)}{3!}(1-0)^3$$

for some $c_1 \in (-1,0)$ and $c_2 \in (0,1)$. If we subtract the first from the second, we obtain:

$$f(1) - f(-1) = 2f'(0) + \frac{f'''(c_2)}{3!} + \frac{f'''(c_1)}{3!} = \frac{1}{6}\left[f'''(c_1) + f'''(c_2)\right].$$

Therefore,

$$f'''(c_1) + f'''(c_2) = 6$$

where $c_1, c_2 \in (-1,1)$ with $c_1 < c_2$. Now we can use the intermediate value theorem for derivatives to obtain.

$$f'''(c) = \left[f'''(c_1) + f'''(c_2)\right]/2 \text{ for some } c \in (c_1, c_2).$$

It follows that

$$2f'''(c) = 6 \text{ for some } c \in (c_1, c_2).$$

Therefore,

$$f'''(c) = 3 \text{ for some } c \in (-1,1).$$

CHAPTER

5 Integration

5.1 The Riemann Integral

In this section, we shall define the Riemann integral of a bounded function defined on a closed interval of the form $[a,b]$, where $a,b \in \mathbb{R}$ with $a<b$. We will also establish all the standard results associated with this integral.

5.1D1 Definition: Partition

Let $[a,b]$ be an interval in $\mathbb{R}$. A **partition** P of $[a,b]$ is a finite set $\{x_0, x_1, ..., x_n\}$ of points in $[a,b]$ satisfying

$$a = x_0 < x_1 < ... < x_n = b, \text{ where } n \in \mathbb{N}.$$

We shall denote the set of all partitions of $[a,b]$ by $P[a,b]$. Furthermore, the length of each subinterval $[x_{i-1}, x_i]$ will be denoted by Δx_i, i.e., $\Delta x_i = x_i - x_{i-1}$ for $i = 1,2,...,n$.

5.1D2 Definition: Upper and Lower Sums

Suppose $f:[a,b] \to \mathbb{R}$ is bounded and $P \in P[a,b]$. Let

$$m_i = \inf\{f(t) | t \in [x_{i-1}, x_i]\} \text{ and } M_i = \sup\{f(t) | t \in [x_{i-1}, x_i]\}$$

for $i = 1,2,...,n$. The **upper sum of *f*** with respect to the partition P, denoted by $U(P,f)$, is defined to be

$$U(P,f) = \sum_{i=1}^{n} M_i \Delta x_i.$$

The **lower sum of *f*** with respect to the partition P, denoted by $L(P,f)$, is defined to be

$$L(P,f) = \sum_{i=1}^{n} m_i \Delta x_i.$$

Since $m_i \le M_i$ for all $i = 1,2,...,n,$ we must have

$$L(P,f) \le U(P,f) \text{ for any } P \in P[a,b].$$

Note: Since *f* is bounded, there exist $m, M \in \mathbb{R}$ such that $m \le f(x) \le M$ for all $x \in [a,b]$ and

$$m(b-a) = \sum_{i=1}^{n} m\Delta x_i \le L(P,f) = \sum_{i=1}^{n} m_i \Delta x_i$$
$$\le U(P,f) = \sum_{i=1}^{n} M_i \Delta x_i \le \sum_{i=1}^{n} M \Delta x_i = M(b-a)$$

for any $P \in P[a,b]$. Therefore, the sets

$$\{L(P,f) | P \in P[a,b]\} \text{ and } \{U(P,f) | P \in P[a,b]\}$$

are bounded.

5.1D3 Definition: Upper and Lower Integrals

Suppose $f:[a,b] \to \mathbb{R}$ is bounded. Then the **Riemann Upper Integral of *f*** on $[a,b]$, denoted by $\overline{\int_a^b} f$, is defined to be

$$\overline{\int_a^b} f = \inf\{U(P,f) | P \in P[a,b]\}.$$

The **Riemann Lower Integral of *f*** on $[a,b]$, denoted by $\underline{\int_a^b} f$, is defined to be

$$\underline{\int_a^b} f = \sup\{L(P,f) | P \in P[a,b]\}.$$

5.1D4 Definition:

Suppose $P, P^* \in P[a,b]$. Then P^* is said to be a **refinement** of *P* if $P \subset P^*$. The **common refinement** of *P* and P^* is $P \cup P^*$ if $P \ne P^*$.

5.1T1 Theorem:

If $f:[a,b] \to \mathbb{R}$ is bounded, $P, P^* \in P[a,b]$, and P^* is a **refinement** of *P*, then

$$L(P,f) \le L(P^*,f) \le U(P^*,f) \le U(P,f).$$

Proof: Suppose P^* contains one more point than *P*, and let $x^* \in (P^* \setminus P)$ with $x_{j-1} < x^* < x_j$ for some $j \in \{1,2,...,n\}$ where $P = \{x_0, x_1, ..., x_n\}, n \in \mathbb{N}$. Let $m_{j1} = \inf\{f(t) | t \in [x_{j-1}, x^*]\}$, $M_{j1} = \sup\{f(t) | t \in [x_{j-1}, x^*]\}$, $m_{j2} = \inf\{f(t) | t \in [x^*, x_j]\}$, and $M_{j2} = \sup\{f(t) | t \in [x^*, x_j]\}$. Then

$$L(P,f) = \sum_{i=1}^{n} m_i \Delta x_i = \sum_{i=1}^{j-1} m_i \Delta x_i + m_j \Delta x_j + \sum_{i=j+1}^{n} m_i \Delta x_i$$

$$= \sum_{i=1}^{j-1} m_i \Delta x_i + m_j \left(x^* - x_{j-1}\right) + m_j \left(x_j - x^*\right) + \sum_{i=j+1}^{n} m_i \Delta x_i$$

$$\le \sum_{i=1}^{j-1} m_i \Delta x_i + m_{j1} \left(x^* - x_{j-1}\right) + m_{j2} \left(x_j - x^*\right) + \sum_{i=j+1}^{n} m_i \Delta x_i$$

$$= L\left(P^*, f\right) \le U\left(P^*, f\right)$$

$$= \sum_{i=1}^{j-1} M_i \Delta x_i + M_{j1} \left(x^* - x_{j-1}\right) + M_{j2} \left(x_j - x^*\right) + \sum_{i=j+1}^{n} M_i \Delta x_i$$

$$\le \sum_{i=1}^{j-1} M_i \Delta x_i + M_j \left(x^* - x_{j-1}\right) + M_j \left(x_j - x^*\right) + \sum_{i=j+1}^{n} M_i \Delta x_i$$

$$= \sum_{i=1}^{j-1} M_i \Delta x_i + M_j \Delta x_j + \sum_{i=j+1}^{n} M_i \Delta x_i$$

$$= \sum_{i=1}^{n} M_i \Delta x_i = U\left(P, f\right).$$

If P^* contains k more points than P, then we use this argument k times to obtain the result.

5.1T2 Theorem:

Suppose $f : [a,b] \to \mathbb{R}$ is bounded. Then $\underline{\int_a^b} f \le \overline{\int_a^b} f$.

Proof: Let $P_1, P_2 \in P[a,b]$ and $P^* = P_1 \cup P_2$, where $P_1 \ne P_2$. Since P^* is a refinement of both P_1 and P_2, we use (5.1T1) to obtain

$$L\left(P_1, f\right) \le L\left(P^*, f\right) \le U\left(P^*, f\right) \le U\left(P_2, f\right).$$

Taking the infimum over P_2, we obtain

$$L\left(P_1, f\right) \le \overline{\int_a^b} f.$$

Now take the supremum over P_1 to obtain the result.

5.1D5 Definition:

Let $f : [a,b] \to \mathbb{R}$ be bounded. Then f is said to be **(Riemann) integrable** on $[a,b]$ if

$$\overline{\int_a^b} f = \underline{\int_a^b} f.$$

The common value is denoted by

$$\int_a^b f \text{ or } \int_a^b f(x)\,dx$$

and is called the **(Riemann) integral** of f over $[a,b]$. The set of all (Riemann) integrable functions is denoted by $\Re[a,b]$. Furthermore, we define

$$\int_b^a f = -\int_a^b f.$$

5.1E1 Example: Define $f:\mathbb{R}\to\mathbb{R}$ by

$$f(x)=\begin{cases}0, & x\in\mathbb{Q}\\ 1, & x\notin\mathbb{Q}\end{cases}$$

and let $P\in P[a,b]$. Then $m_i=0$ and $M_i=1$ for $i=1,2,...,n$. Therefore,

$$L(P,f)=0 \text{ and } U(P,f)=1$$

for any $P\in P[a,b]$, and it follows that

$$\underline{\int_a^b} f=0,\ \overline{\int_a^b} f=b-a, \text{ and } f\notin\Re[a,b].$$

5.1E2 Example: Define $f:[0,2]\to\mathbb{R}$ by

$$f(x)=\begin{cases}1, & x\in[0,1)\\ 2, & x\in[1,2]\end{cases}$$

and let $P\in P[0,2]$ with $x_{k-1}<1\le x_k$ for some $k\in\{1,2,...,n\}$. Then

$$m_i=\begin{cases}1, & i=1,2,...,k\\ 2, & i=k+1,...,n\end{cases} \text{ and } M_i=\begin{cases}1, & i=1,...,k-1\\ 2, & i=k,...,n.\end{cases}$$

Therefore,

$$L(P,f)=\sum_{i=1}^{k}1\Delta x_i+\sum_{i=k+1}^{n}2\Delta x_i=4-x_k$$

and

$$U(P,f)=\sum_{i=1}^{k-1}1\Delta x_i+\sum_{i=k}^{n}2\Delta x_i=4-x_{k-1}.$$

Since $4-x_k\le 4-1<4-x_{k-1}$,

$$L(P,f)\le 3<U(P,f) \text{ for all } P\in P[0,2].$$

Hence,

$$\underline{\int_0^2} f\le 3\le \overline{\int_0^2} f.$$

If f is integrable, then $\int_0^2 f=3$. We will now show that f is integrable.

Let $\varepsilon>0$ be given and choose $P\in P[0,2]$ such that $\Delta x_k<\varepsilon$, where k is as defined above. Then

$$U(P,f)-L(P,f)=x_k-x_{k-1}.$$

Therefore,

$$U(P,f)=L(P,f)+\Delta x_k<L(P,f)+\varepsilon$$

and

$$\overline{\int_0^2} f\le U(P,f)<L(P,f)+\varepsilon\le\underline{\int_0^2} f+\varepsilon.$$

Since $\varepsilon > 0$ was arbitrary,

$$\overline{\int_0^2} f \le \underline{\int_0^2} f.$$

Thus

$$\overline{\int_0^2} f = \underline{\int_0^2} f, \; f \in \Re[a,b], \text{ and } \int_0^2 f = 3.$$

Note: As can be seen from the above example, it is not easy to show that a function is integrable, nor is it easy to compute the integral. We need further results to show integrability and to compute the integrals.

5.1T3 Theorem:

Suppose $f:[a,b] \to \mathbb{R}$ is bounded. Then $f \in \Re[a,b]$ if and only if for every $\varepsilon > 0$ there exists $P \in P[a,b]$ such that

$$U(P,f) - L(P,f) < \varepsilon.$$

Proof: Suppose $f \in \Re[a,b]$ and let $\varepsilon > 0$ be given. Then there exist $P_1, P_2 \in P[a,b]$ such that

$$U(P_1,f) - \int_a^b f < \varepsilon/2 \text{ and } \int_a^b f - L(P_2,f) < \varepsilon/2.$$

Let $P = P_1 \cup P_2$. Since P is a refinement of both P_1 and P_2,

$$U(P,f) \le U(P_1,f) < \int_a^b f + \varepsilon/2$$
$$< L(P_2,f) + \varepsilon \le L(P,f) + \varepsilon.$$

Therefore,

$$U(P,f) - L(P,f) < \varepsilon.$$

Suppose for every $\varepsilon > 0$ there exists $P \in P[a,b]$ such that

$$U(P,f) - L(P,f) < \varepsilon.$$

Then since $L(P,f) \le \underline{\int_a^b} f \le \overline{\int_a^b} f \le U(P,f)$,

$$0 \le \overline{\int_a^b} f - \underline{\int_a^b} f \le U(P,f) - L(P,f) < \varepsilon.$$

But $\varepsilon > 0$ was arbitrary. Therefore,

$$\overline{\int_a^b} f = \underline{\int_a^b} f \text{ and } f \in \Re[a,b].$$

5.1T4 Theorem:

Suppose $f:[a,b]\to\mathbb{R}$ is bounded and continuous on $[a,b]\setminus\{t_1,t_2,...,t_k\}$ and discontinuous at each point in $\{t_1,t_2,...,t_k\}$ where $k\in\mathbb{N}$ and the discontinuities satisfy $a\le t_1<t_2<...<t_k\le b$. Then $f\in\Re[a,b]$.

Proof: Let $\varepsilon>0$ be given, $M=\sup\{|f(x)|\,|\,x\in[a,b]\}$, and $\varepsilon'=\dfrac{\varepsilon}{2M+b-a}$. Then there exist $a_1,b_1,a_2,b_2,...,a_k,b_k\in\mathbb{R}$ such that

$$a_1<b_1<a_2<b_2<...<a_k<b_k,\ \ a_i<t_i<b_i \text{ for } i=1,2,...,k$$

and

$$\sum_{i=1}^{k}(b_i-a_i)<\varepsilon'.$$

Let $K=[a,b]\setminus\bigcup_{i=1}^{k}(a_i,b_i)$. Then K is compact and f is uniformly continuous on K because it is continuous on K. Therefore, there exists $\delta>0$ such that

$$|f(x)-f(y)|<\varepsilon' \text{ whenever } x,y\in K \text{ and } |x-y|<\delta.$$

Let $P\in P[a,b]$ satisfy the following conditions:

(a) $\{a_1,b_1,a_2,b_2,...,a_k,b_k\}\cap[a,b]\subseteq P$ and $P\cap\bigcup_{i=1}^{k}(a_i,b_i)=\phi$.

(b) $\Delta x_i<\delta$ if $x_{i-1}\notin\{a_1,a_2,...,a_k\}$ for $i=1,2,...,n$.

Then

$$M_i-m_i\le 2M \text{ for } i=1,2,...,n$$

and

$$M_i-m_i\le\varepsilon \text{ if } x_{i-1}\notin\{a_1,a_2,...,a_k\} \text{ for } i=1,2,...,n.$$

It follows that

$$\begin{aligned} U(P,f)-L(P,f) &\le \sum_{i=1}^{k}(M_i-m_i)(b_i-a_i)+\sum_{\substack{i=1 \text{ and}\\ x_i\ne b_j \text{ for any } j}}^{n}(M_i-m_i)\Delta x_i \\ &\le \sum_{i=1}^{k}2M(b_i-a_i)+\sum_{\substack{i=1 \text{ and}\\ x_i\ne b_j \text{ for any } j}}^{n}\varepsilon'\Delta x_i \\ &<(2M+b-a)\varepsilon'=\varepsilon. \end{aligned}$$

Hence, $f\in\Re[a,b]$ by (5.1T3).

5.1T5 Theorem: Mean Value Theorem for Integrals

If $f:[a,b]\to\mathbb{R}$ is continuous, then there exists $c\in[a,b]$ such that

$$\int_a^b f=(b-a)f(c).$$

Proof: Since *f* is continuous on $[a,b]$,

$$f(s)=\min_{[a,b]} f = m \text{ and } f(t)=\max_{[a,b]} f = M$$

where $m, M \in \mathbb{R}$ and $s,t \in [a,b]$. Then

$$m(b-a)=\int_a^b m \le \int_a^b f \le \int_a^b M = M(b-a),$$

because for all $x \in [a,b]$, $m \le f(x) \le M$ and

$$f(s)=m\le \frac{1}{b-a}\int_a^b f \le M = f(t).$$

Thus, by the Intermediate Value Theorem, there exists *c* between *s* and *t* such that

$$f(c)=\frac{1}{b-a}\int_a^b f.$$

Hence result.

Note: If $f:[a,b]\to\mathbb{R}$ is bounded and $f\in\mathfrak{R}[a,b]$, then $\int_a^b f=(b-a)\lambda$ for some $\lambda\in[m,M]$, where $m=\inf\limits_{[a,b]} f$ and $M=\sup\limits_{[a,b]} f$. We can show this by a similar argument as above.

Note: We have shown that all polynomials and rational functions are continuous. Therefore, they are integrable on any interval of the form $[a,b]$ of their domains.

5.1E3 Example: Define $f:[0,1]\to\mathbb{R}$ as $f(x)=x^3$, $x\in[0,1]$. Then *f* is continuous and must be integrable on $[0,1]$ by (5.1T5). Define $P_n\in P[0,1]$ as

$$x_i=\frac{i}{n},\ i=0,1,\ldots,n. \text{ Let } h=1/n.$$

Then

$$m_i = x_{i-1}{}^3=(i-1)^3h^3,\ M_i=x_i{}^3=i^3h^3,\ \Delta x_i = h \text{ for } i=0,1,\ldots,n.$$

Therefore,

$$\begin{aligned} L(P_n,f)&=\sum_{i=1}^n m_i\Delta x_i=\sum_{i=1}^n (i-1)^3h^4\\ &=\left[0^3+1^3+\ldots+(n-1)^3\right]h^4\\ &=\frac{(n-1)^2n^2}{4}\frac{1}{n^4}\end{aligned}$$

and

$$U(P_n, f) = \sum_{i=1}^{n} M_i \Delta x_i = \sum_{i=1}^{n} i^3 h^4$$
$$= \left[1^3 + 2^3 + \ldots + n^3\right] h^4$$
$$= \frac{n^2 (n+1)^2}{4} \frac{1}{n^4}.$$

But

$$L(P_n, f) \leq \underline{\int_0^1} f \Rightarrow \frac{1}{4} = \lim_{n \to \infty} L(P_n, f) \leq \underline{\int_0^1} f$$

and

$$U(P_n, f) \geq \overline{\int_0^1} f \Rightarrow \frac{1}{4} = \lim_{n \to \infty} U(P_n, f) \geq \overline{\int_0^1} f.$$

This implies that

$$\overline{\int_0^1} f \leq \frac{1}{4} \leq \underline{\int_0^1} f.$$

Therefore,

$$\underline{\int_0^1} f = \overline{\int_0^1} f = \int_0^1 f = \frac{1}{4}.$$

5.1A1 Assignment: Define $f : [0,2] \to \mathbb{R}$ by $f(x) = x^2$, $x \in [0,2]$. Use the ideas in the above example to evaluate $\int_0^2 f$.

5.1E4 Example: Suppose $f : [a,b] \to \mathbb{R}$ is continuous and $f(x) \geq 0$ for all $x \in [a,b]$. Prove that if $\int_a^b f = 0$, then $f(x) = 0$ for all $x \in [a,b]$.

Suppose $f(c) > 0$ for some $c \in [a,b]$. Since *f* is continuous at *c*, there exists $h > 0$ such that

$$f(x) > \frac{f(c)}{2} \text{ for all } x \in [d, d+h] \subseteq [a,b].$$

Choose $P \in P[a,b]$ such that $d, (d+h) \in P$. Then

$$\int_a^b f \text{ exists and } \int_a^b f \geq L(P,f) \geq \frac{hf(c)}{2} > 0.$$

This is a contradiction. Hence result.

5.1A2 Assignment:

(a) Suppose $f,g:[a,b]\to\mathbb{R}$ are bounded and $f,g\in\Re[a,b]$ with $f(x)\le g(x)$ for all $x\in[a,b]$. Prove that $\int_a^b f\le\int_a^b g$.

(b) Suppose $f:[a,b]\to\mathbb{R}$ is increasing on $[a,b]$, $n\in\mathbb{N}$, $h=\dfrac{b-a}{n}$, and $P_n=\{x_0,x_1,\ldots,x_n\}$ where $x_i=a+ih$ for $i=0,1,\ldots,n$. Prove that $0\le U(P_n,f)-\int_a^b f\le\dfrac{b-a}{n}[f(b)-f(a)]$ and $\lim\limits_{n\to\infty}U(P_n,f)=\int_a^b f$.

(c) Use (b) to evaluate $\int_a^b x\,dx$ and $\int_a^b x^3dx$.

(d) Suppose $f:[a,b]\to\mathbb{R}$ is bounded, $f\in\Re[a,b]$, and $f(x)\ge 0$ for all $x\in[a,b]$. Prove that if $f(r)=0$ for all $r\in\mathbb{Q}\cap[a,b]$, then $\int_a^b f=0$.

5.1T6 Theorem:

Suppose $f:[a,b]\to\mathbb{R}$ is bounded and monotonic on $[a,b]$. Then $f\in\Re[a,b]$.

Proof: Without loss of generality, we may assume that *f* is decreasing. Let $\varepsilon>0$ be given and let $n\in\mathbb{N}$ satisfy $n>\dfrac{(b-a)[f(a)-f(b)]}{\varepsilon}$, $h=\dfrac{b-a}{n}$, and $x_i=a+ih$ for $i=0,1,\ldots,n$. Then $P=\{x_0,x_1,\ldots x_n\}\in P[a,b]$ and *P* satisfies $\Delta x_i=h$ for $i=0,1,\ldots,n$. Since *f* is decreasing on $[a,b]$, $m_i=f(x_i)$ and $M_i=f(x_{i-1})$ for $i=0,1,\ldots,n$. Therefore,

$$\begin{aligned}U(P,f)-L(P,f)&=\sum_{i=1}^{n}(M_i-m_i)\Delta x_i\\&=\sum_{i=1}^{n}[f(x_{i-1})-f(x_i)]h\\&=\frac{b-a}{n}[f(a)-f(b)]<\varepsilon.\end{aligned}$$

Hence, $f\in\Re[a,b]$ by (5.1T3).

5.1T7 Theorem:

Suppose $f:[a,b]\to\mathbb{R}$ is bounded, $f\in\Re[a,b]$, and $g:[c,d]\to\mathbb{R}$ is continuous where $f([a,b])\subseteq[c,d]$. Then

$$h=g\circ f\in\Re[a,b].$$

Proof: Since g is continuous on $[c,d]$, it is bounded and uniformly continuous on $[c,d]$. Let $\varepsilon > 0$ be given, $K = \max\{|g(t)| \mid t \in [c,d]\}$, and $\varepsilon' = \dfrac{\varepsilon}{2K + b - a}$. Then there exists $\delta > 0$ such that $\delta < \varepsilon'$ and

$$|g(s) - g(t)| < \varepsilon' \text{ whenever } s,t \in [c,d] \text{ and } |s - t| < \delta.$$

Since $f \in \Re[a,b]$, there exists $P \in P[a,b]$ such that

$$U(P,f) - L(P,f) < \delta^2.$$

Define

$$m_i^* = \inf\{h(t) \mid t \in [x_{i-1}, x_i]\} \text{ and } M_i^* = \sup\{h(t) \mid t \in [x_{i-1}, x_i]\} \text{ for } i = 1,2,\dots,n.$$

Let

$$A = \{i \mid M_i - m_i < \delta\} \text{ and } B = \{i \mid M_i - m_i \geq \delta\}.$$

Then

$$A \cap B = \phi \text{ and } A \cup B = \{1,2,\dots,n\}.$$

Therefore,

$$M_i^* - m_i^* = \sup\{h(s) - h(t) \mid s,t \in [x_{i-1}, x_i]\} \leq \varepsilon' \text{ if } i \in A$$

because $|f(s) - f(t)| \leq M_i - m_i$. If $i \in B$, then

$$M_i^* - m_i^* \leq 2K.$$

Since $M_i - m_i \geq \delta$ for $i \in B$,

$$\begin{aligned}\sum_{i \in B} \Delta x_i &\leq \sum_{i \in B} \frac{M_i - m_i}{\delta} \Delta x_i \\ &= \frac{U(P,f) - L(P,f)}{\delta} \\ &< \delta < \varepsilon'.\end{aligned}$$

Therefore,

$$\begin{aligned}U(P,h) - L(P,h) &= \sum_{i \in A} (M_i^* - m_i^*) \Delta x_i + \sum_{i \in B} (M_i^* - m_i^*) \Delta x_i \\ &\leq \sum_{i \in B} \varepsilon' \Delta x_i + \sum_{i \in B} 2K \Delta x_i \\ &< \varepsilon'(b-a) + 2K\varepsilon' = \varepsilon.\end{aligned}$$

The result follows by (5.1T3).

5.1T7C Corollary:

Suppose $f:[a,b]\to\mathbb{R}$ is bounded and $f\in\Re[a,b]$. Then $|f|, f^n\in\Re[a,b]$ where $n\in\mathbb{N}$. The functions $g_1(x)=|x|$, $x\in\mathbb{R}$ and $g_2(x)=x^n$, $x\in\mathbb{R}$ are clearly continuous. Therefore,

$$g_1\circ f=|f|\in\Re[a,b] \text{ and } g_2\circ f=f^n\in\Re[a,b].$$

5.1E5 Example: Suppose $f:[a,b]\to\mathbb{R}$ is bounded and $f^n\in\Re[a,b]$, where $n\in\mathbb{N}$. Show that $f\in\Re[a,b]$ if *n* is odd, or *n* is even and $f([a,b])\subseteq(0,\infty)$.

Suppose *n* is odd. Then $\mathbb{R}\to\mathbb{R}$ defined by $g_1(x)=x^{1/n}$, $x\in\mathbb{R}$ is clearly continuous. Therefore, $g_1\circ f^n=f\in\Re[a,b]$.

Suppose *n* is even and $f([a,b])\subseteq(0,\infty)$. Then we can define a continuous function $g_2(x)=x^{1/n}$, $x\in(0,\infty)$. In this case, $g_2\circ f^n=f\in\Re[a,b]$.

5.1A3 Assignment: Suppose $f,g:[a,b]\to\mathbb{R}$ are bounded and $f,g\in\Re[a,b]$. Prove that

(a) $\left|\int_a^b fg\right|\le\left[\left(\int_a^b f^2\right)\left(\int_a^b g^2\right)\right]^{1/2}$ (Assume that $fg\in\Re[a,b]$. This will be established in (5.2T1.) This is the **Cauchy-Bunyakovsky-Schwarz** inequality for integrals.

Hint: Expand $\int_a^b (tf+g)^2$ where $t\in\mathbb{R}$ as a quadratic in *t* and use the discriminant from the quadratic formula.

(b) $\left[\int_a^b (f+g)^2\right]^{1/2}\le\left[\int_a^b f^2\right]^{1/2}+\left[\int_a^b g^2\right]^{1/2}$. This is **Minkowski**'s inequality for integrals. Hint: Expand the integral on the left and use (a).

5.2 Properties of the Riemann Integral

In this section, we shall derive some of the basic properties of the Riemann integral of a bounded function defined on a closed interval of the form $[a,b]$, where $a<b$.

5.2T1 Theorem:

Suppose $f,g:[a,b]\to\mathbb{R}$ are bounded, $f,g\in\Re[a,b]$, and $c\in\mathbb{R}$. Then

(a) $f,g\in\Re[a,b]$ and $\int_a^b(f+g)=\int_a^b f+\int_a^b g$.

(b) $cf\in\Re[a,b]$ and $\int_a^b cf=c\int_a^b f$.

(c) $fg\in\Re[a,b]$.

Proof:

(a) Let $P \in P[a,b]$. Since $\sup_E(f+g) \le \sup_E f + \sup_E g$ for any $E \subseteq [a,b]$, it follows that

$$U(P,f+g) \le U(P,f)+U(P,g).$$

Let $\varepsilon > 0$ be given. There exist $P_1, P_2 \in P[a,b]$ such that

$$U(P_1,f) < \int_a^b f + \varepsilon/2 \text{ and } U(P_2,g) < \int_a^b g + \varepsilon/2$$

because $f, g \in \Re[a,b]$. Let $Q = P_1 \cup P_2$. Since Q is a refinement of both P_1 and P_2,

$$\begin{aligned} U(Q,f+g) &\le U(P_1,f)+U(P_2,g) \\ &< \int_a^b f + \int_a^b g + \varepsilon. \end{aligned}$$

It follows that

$$\overline{\int_a^b}(f+g) < \int_a^b f + \int_a^b g + \varepsilon.$$

Since $\varepsilon > 0$ was arbitrary,

$$\overline{\int_a^b}(f+g) \le \int_a^b f + \int_a^b g.$$

With a similar argument, we can show that

$$\underline{\int_a^b}(f+g) \ge \int_a^b f + \int_a^b g.$$

But $\underline{\int_a^b} \le \overline{\int_a^b}$. Therefore, $\underline{\int_a^b}(f+g) = \overline{\int_a^b}(f+g) = \int_a^b f + \int_a^b g$.

(b) If $c = 0$, then the result is trivially true. Suppose $c > 0$. Then for any $P \in P[a,b]$,

$$U(P,cf) = cU(P,f) \text{ and } L(P,cf) = cL(P,f).$$

Hence,

$$\overline{\int_a^b} cf = c\overline{\int_a^b} f = c\underline{\int_a^b} f = \underline{\int_a^b} cf.$$

Therefore,

$$cf \in \Re[a,b] \text{ and } \int_a^b cf = c\int_a^b f.$$

If $c < 0$, then for any $P \in P[a,b]$,

$$U(P,cf) = cL(P,f) \text{ and } L(P,cf) = cU(P,f).$$

Hence,

$$\overline{\int_a^b} cf = c\underline{\int_a^b} f = c\overline{\int_a^b} f = \underline{\int_a^b} cf.$$

Therefore, $cf \in \Re[a,b]$ and $\int_a^b cf = c\int_a^b f$.

(c) Since $g \in \Re[a,b]$, $(-1)g \in \Re[a,b]$ and $f - g = f + (-1)g \in \Re[a,b]$. By (5.1T7C),

$$(f+g)^2,(f-g)^2 \in \Re[a,b].$$

Therefore, by (a) and (b), $\dfrac{(f+g)^2-(f-g)^2}{4} = fg \in \Re[a,b]$.

5.2T2 Theorem:

If $f:[a,b]\to\mathbb{R}$ is bounded and $f\in\Re[a,b]$, then $\left|\int_a^b f\right| \le \int_a^b |f|$.

Proof: We have already shown that $|f|\in\Re[a,b]$. Let $c=\pm 1$ be such that $\left|\int_a^b f\right| = c\int_a^b f$. Then

$$\left|\int_a^b f\right| = c\int_a^b f = \int_a^b cf \le \int_a^b |f|$$

because $cf(x) = \pm f(x) \le |f(x)|$ for all $x\in[a,b]$.

5.2E1 Example: Let $f:[a,b]\to\mathbb{R}$ be continuous and nonnegative on $[a,b]$. Prove that $f(c) = \left(\dfrac{1}{b-a}\int_a^b f^2\right)^{1/2}$ for some $c\in[a,b]$.

Since f is continuous, $f\in\Re[a,b]$. Therefore, $f^2\in\Re[a,b]$ and by the MVT for integrals, there exists $c\in[a,b]$ such that

$$f^2(c) = \frac{1}{b-a}\int_a^b f^2.$$

Since f is nonnegative, we obtain the result.

5.2E2 Example: Suppose $f,g:[a,b]\to\mathbb{R}$ are bounded, f is continuous, g is nonnegative on $[a,b]$, and $g\in\Re[a,b]$. Prove that $\int_a^b fg = f(c)\int_a^b g$ for some $c\in[a,b]$.

If $g\equiv 0$, then the result holds for any $c\in[a,b]$. Assume that $g(x)>0$ for some x. Then $\int_a^b g>0$. Since f is continuous on $[a,b]$,

$$f(s) = \min_{[a,b]} f = m \text{ and } f(t) = \max_{[a,b]} f = M$$

where $m,M\in\mathbb{R}$ and $s,t\in[a,b]$. By (5.2T2(c)),

$$f,g\in\Re[a,b] \Rightarrow fg\in\Re[a,b].$$

For all $x\in[a,b]$,

$$m\le f(x)\le M \Rightarrow mg(x)\le f(x)g(x)\le Mg(x)$$

$$\Rightarrow m\int_a^b g \le \int_a^b fg \le M\int_a^b g \Rightarrow m \le \frac{\int_a^b fg}{\int_a^b g} \le M.$$

Since $f(s)=m$ and $f(t)=M$, there exists $c\in[a,b]$ such that $f(c)=\dfrac{\int_a^b fg}{\int_a^b g}$. Hence result.

5.2T3 Theorem:

Suppose $f:[a,b]\to\mathbb{R}$ is bounded and $c\in(a,b)$. Then $f\in\Re[a,b]$ if and only if $f\in\Re[a,c]$ and $f\in\Re[c,b]$. In this case, $\int_a^b f=\int_a^c f+\int_c^b f$.

Proof: Suppose $P_1\in P[a,c]$ and $P_2\in P[c,b]$. Then $P_1\cup P_2=P\in P[a,b]$. Conversely, if $P\in P[a,b]$ with $c\in P$, then $P=P_1\cup P_2$ where $P_1\in P[a,c]$ and $P_2\in P[c,b]$. For such a P,

$$\begin{aligned}U(P,f)&=U(P_1,f)+U(P_2,f)\\&\geq\overline{\int_a^c} f+\overline{\int_c^b} f.\end{aligned}$$

Now let $Q\in P[a,b]$. Then $P=Q\cup\{c\}$ is a refinement of Q containing c. Therefore,

$$U(Q,f)\geq U(P,f)\geq\overline{\int_a^c} f+\overline{\int_c^b} f.$$

Taking the infimum of all $Q\in P[a,b]$, we obtain

$$\overline{\int_a^b} f\geq\overline{\int_a^c} f+\overline{\int_c^b} f.$$

Suppose $\varepsilon>0$ is given. Then there exist $P_1\in P[a,c]$ and $P_2\in P[c,b]$ such that

$$U(P_1,f)<\overline{\int_a^c} f+\varepsilon/2 \text{ and } U(P_2,f)<\overline{\int_c^b} f+\varepsilon/2.$$

Let $P_1\cup P_2=P\in P[a,b]$. Then

$$\begin{aligned}\overline{\int_a^b} f\leq U(P,f)&=U(P_1,f)+U(P_2,f)\\&<\overline{\int_a^c} f+\overline{\int_c^b} f+\varepsilon.\end{aligned}$$

Since $\varepsilon>0$ was arbitrary, it follows that

$$\overline{\int_a^b} f\leq\overline{\int_a^c} f+\overline{\int_c^b} f.$$

Therefore,

$$\overline{\int_a^b} f=\overline{\int_a^c} f+\overline{\int_c^b} f.$$

Using a similar argument, we can show that

$$\underline{\int_a^b} f=\underline{\int_a^c} f+\underline{\int_c^b} f.$$

If $f \in \Re[a,b]$, then $\underline{\int_a^b} f = \overline{\int_a^b} f = \int_a^b f$. Therefore,

$$\int_a^b f = \underline{\int_a^c} f + \underline{\int_c^b} f = \overline{\int_a^c} f + \overline{\int_c^b} f.$$

Since $\underline{\int_a^c} f \le \overline{\int_a^c} f$ and $\underline{\int_c^b} f \le \overline{\int_c^b} f$, it follows that

$$\underline{\int_a^c} f = \overline{\int_a^c} f \text{ and } \underline{\int_c^b} f = \overline{\int_c^b} f.$$

Hence,

$$f \in \Re[a,c],\ f \in \Re[c,b], \text{ and } \int_a^b f = \int_a^c f + \int_c^b f.$$

Conversely, if $f \in \Re[a,c]$ and $f \in \Re[c,b]$, then

$$\int_a^c f + \int_c^b f = \underline{\int_a^b} f \le \overline{\int_a^b} f = \int_a^c f + \int_c^b f.$$

Thus

$$f \in \Re[a,b] \text{ and } \int_a^b f = \int_a^c f + \int_c^b f.$$

5.2D1 Definition: Riemann Sum

Suppose $f:[a,b] \to \mathbb{R}$ is bounded and $P \in P[a,b]$. Choose $t_i \in [x_{i-1}, x_i]$ for each $i = 1,2,\ldots,n$. The sum

$$S(P,f) = \sum_{i=1}^{n} f(t_i)\Delta x_i$$

is called a **Riemann sum** of *f* with respect to *P* and the choice of points $\alpha(P) = \{t_1, t_2, \ldots, t_n\}$. The **norm** or mesh of *P*, denoted by $\|P\|$, is given by $\|P\| = \max\{\Delta x_i | i = 1,2,\ldots,n\}$.

5.2D2 Definition:

Suppose $f:[a,b] \to \mathbb{R}$ is bounded and $L \in \mathbb{R}$. Then $\lim_{\|P\| \to 0} S(P,f) = L$ if for every $\varepsilon > 0$ there exist $\delta > 0$ such that

$$|S(P,f) - L| = \left| \sum_{i=1}^{n} f(t_i)\Delta x_i - L \right| < \varepsilon$$

for all $P \in P[a,b]$ with $\|P\| < \delta$ and all choices $\alpha(P)$.

5.2T4 Theorem:

Suppose $f:[a,b] \to \mathbb{R}$ is bounded and $L \in \mathbb{R}$. If $\lim_{\|P\| \to 0} S(P,f) = L$, then $f \in \Re[a,b]$ and $L = \int_a^b f$. Conversely, if $f \in \Re[a,b]$, then $\lim_{\|P\| \to 0} S(P,f)$ exists and $\lim_{\|P\| \to 0} S(P,f) = \int_a^b f$.

Proof: Suppose $\lim_{\|P\|\to 0} S(P,f) = \int_a^b f$. Let $\varepsilon > 0$ be given and $\varepsilon' = \dfrac{\varepsilon}{1+b-a}$. Then there exist $\delta > 0$ such that

$$\left|S(P,f) - L\right| = \left|\sum_{i=1}^{n} f(t_i)\Delta x_i - L\right| < \varepsilon'$$

for all $P \in P[a,b]$ with $\|P\| < \delta$ and all choices $\alpha(P)$. Then for each $i = 1,2,\dots,n,$ there exists $t_i \in [x_{i-1}, x_i]$ such that

$$f(t_i) > M_i - \varepsilon'.$$

Thus

$$U(P,f) = \sum_{i=1}^{n} M_i \Delta x_i < \sum_{i=1}^{n} f(t_i)\Delta x_i + \sum_{i=1}^{n} \varepsilon' \Delta x_i$$
$$< L + \varepsilon' + \varepsilon'(b-a) = L + \varepsilon.$$

Similarly,

$$L(P,f) > L - \varepsilon.$$

Therefore,

$$\overline{\int_a^b} f \le U(P,f) < L + \varepsilon \text{ and } \underline{\int_a^b} f \ge L(P,f) > L - \varepsilon.$$

Since $\varepsilon > 0$ was arbitrary,

$$\overline{\int_a^b} f \le L \text{ and } \underline{\int_a^b} f \ge L.$$

Hence,

$$f \in \Re[a,b] \text{ and } \int_a^b f = L$$

because $\overline{\int_a^b} f \ge \underline{\int_a^b} f.$

Conversely, suppose $f \in \Re[a,b]$ and let $M = \sup_{[a,b]} |f|$. Let $\varepsilon > 0$ be given. Then there exists $Q = \{y_0, y_1, \dots, y_m\} \in P[a,b]$ such that

$$\int_a^b f - \frac{\varepsilon}{3} < L(Q,f) \le U(Q,f) < \int_a^b f + \frac{\varepsilon}{3}.$$

Let $\delta = \dfrac{\varepsilon}{3mM}$ and $P = \{x_0, x_1, \dots, x_n\} \in P[a,b]$ be such that $\|P\| < \delta$. Consider the interval $[x_{i-1}, x_i]$; $i = 1,2,\dots,n$. This interval may or may not contain points from Q. Since Q contains $m+1$ points, there are at most $m-1$ intervals $[x_{i-1}, x_i]$; $i = 1,2,\dots,n$ which contain a point from $\{y_1, \dots, y_{m-1}\}$. Suppose

$$\{y_j, y_{j+1}, \dots, y_{j+p}\} \subset [x_{i-1}, x_i]$$

for some $i \in \{1,2,...,n\}$, $j \in \{1,2,...,m-1\}$, and $p \in \{1,2,...,m-1-j\}$.

Let

$$P_i = \{x_{i-1}, y_j, y_{j+1}, ..., y_{j+p}, x_i\} \in P[x_{i-1}, x_i],$$
$$M'_{j+k} = \sup\{f(x) | x \in [y_{j+k-1}, y_{j+k}]\}; \ k = 1, ..., p,$$
$$M_i^1 = \sup\{f(x) | x \in [x_{i-1}, y_j]\},$$
$$M_i^2 = \sup\{f(x) | x \in [y_{j+p}, x_i]\}.$$

Suppose $t_i \in [x_{i-1}, x_i]$ is arbitrary. Since $|f(t) - f(s)| \leq 2M$ for all $s, t \in [a,b]$,

$$f(t_i) \leq 2M + M'_{j+k}; \ k = 1, ..., p$$

and

$$f(t_i) \leq 2M + M_i^k; \ k = 1, 2.$$

Therefore,

$$\begin{aligned} f(t_i)\Delta x_i &= f(t_i)(y_j - x_{i-1}) + \sum_{k=1}^{p} f(t_i)\Delta y_{j+k} + f(t_i)(x_i - y_{j+p}) \\ &\leq 2M(y_j - x_{i-1} + x_i - y_{j+p}) + M_i^1(y_j - x_{i-1}) \\ &\quad + \sum_{k=1}^{p} M'_{j+k}\Delta y_{j+k} + M_i^2(x_i - y_{j+p}) \\ &\leq 2M\Delta x_i + U(P_i, f) < 2M\delta. + U(P_i, f). \end{aligned}$$

Let $P_i = \{x_{i-1}, x_i\}$ if $[x_{i-1}, x_i] \cap \{y_1, y_2, ..., y_{m-1}\} = \phi$ and let $t_i \in [x_{i-1}, x_i]$ be arbitrary. If $P' = \bigcup_{i=1}^{n} P_i$, then P' is a refinement of both P and Q. Since at most $m-1$ intervals $[x_{i-1}, x_i]; \ i = 1, 2, ..., n$, contain a point from $\{y_1, ..., y_{m-1}\}$,

$$\begin{aligned} S(P, f) &= \sum_{i=1}^{n} f(t_i)\Delta x_i \\ &< 2M(m-1)\delta + \sum_{i=1}^{n} U(P_i, f) \\ &< \frac{2\varepsilon}{3} + U(P', f) \\ &< \frac{2\varepsilon}{3} + U(Q, f) < \int_a^b f + \varepsilon. \end{aligned}$$

A similar argument shows that $S(P, f) > \int_a^b f - \varepsilon$. Therefore,

$$\left| S(P, f) - \int_a^b f \right| < \varepsilon.$$

The above argument holds for any $P \in P[a,b]$ with $\|P\| < \delta$ and any choice of points $\alpha(P) = \{t_1, t_2, ..., t_n\}$. Therefore,

$$\lim_{\|P\| \to 0} S(P, f) = \int_a^b f.$$

5.2E4 Example: Use the method of Riemann sums to evaluate $\int_a^b x^n dx$, where $n \in \mathbb{N}$; $n \geq 2$.

The function $f(x) = x^n$, $x \in [a,b]$ is continuous. Hence, $f \in \Re[a,b]$ and $\lim_{\|P\| \to 0} S(P, f) = \int_a^b f$. Since this limit exists for any $\alpha(P) = \{t_1, t_2, ..., t_n\}$, let

$$t_i = \sqrt[n]{\frac{x_{i-1}{}^n + x_{i-1}{}^{n-1} x_i + ... + x_{i-1} x_i{}^{n-1} + x_i{}^n}{n+1}}; \; i = 1, 2, ..., n.$$

Then

$$S(P, f) = \sum_{i=1}^{n} \frac{x_{i-1}{}^n + x_{i-1}{}^{n-1} x_i + ... + x_{i-1} x_i{}^{n-1} + x_i{}^n}{n+1} (x_i - x_{i-1})$$

$$= \sum_{i=1}^{n} \frac{x_i{}^{n+1} - x_{i-1}{}^{n+1}}{n+1} = \frac{b^{n+1} - a^{n+1}}{n+1}.$$

Therefore, $\int_a^b x^n dx = \dfrac{b^{n+1} - a^{n+1}}{n+1}$.

5.2A1 Assignment:

(a) Let $f : [0,1] \to \mathbb{R}$ be continuous. Prove that $\lim_{n \to \infty} \left[\frac{1}{n} \sum_{k=1}^{n} f\left(\frac{k}{n}\right) \right] = \int_a^b f$.

(b) Use (a) to evaluate the following limit: $\lim_{n \to \infty} \frac{1}{n^4} \sum_{k=1}^{n} k^3$.

(c) Suppose $f, g, h : [a,b] \to \mathbb{R}$ are bounded and $f(x) \leq g(x) \leq h(x)$ for all $x \in [a,b]$. If $f, h \in \Re[a,b]$ with $\int_a^b f = \int_a^b h = I$, then prove that $g \in \Re[a,b]$ with $\int_a^b g = I$.

5.3 Fundamental Theorem of Calculus

In this section we will prove two important results: the fundamental theorem of calculus and that integration is the inverse operation of differentiation.

5.3D1 Definition: Antiderivative of a Function

Let $f, F : I \to \mathbb{R}$ where *I* is an interval *F* is said to be an **antiderivative** of *f* on *I* if

$F'(x) = f(x)$ for all $x \in I$.

Note: If *F* and *G* are antiderivatives of *f* on *I*, then for some $C \in \mathbb{R}$,

$$F(x) = G(x) + C \text{ for all } x \in I.$$

5.3T1 Theorem: Fundamental Theorem of Calculus

Suppose $f:[a,b] \to \mathbb{R}$ is bounded and $f \in \Re[a,b]$. If F is an antiderivative of f on $[a,b]$, then

$$\int_a^b f = \int_a^b f(x)\,dx = F(b) - F(a).$$

Proof: Let $P = P[a,b]$. Since F is an antiderivative of f, it is differentiable on $[a,b]$. Then by the MVT, there exists $t_i \in (x_{i-1}, x_i)$ such that

$$F(x_i) - F(x_{i-1}) = F'(t_i)\Delta x_i = f(t_i)\Delta x_i$$

for each $i = 1, 2, \ldots, n$. Therefore,

$$\sum_{i=1}^{n} f(t_i)\Delta x_i = \sum_{i=1}^{n}\left[F(x_i) - F(x_{i-1})\right] = F(b) - F(a).$$

Furthermore, $L(P,f) \le \sum_{i=1}^{n} f(t_i)\Delta x_i \le U(P,f)$. Hence,

$$L(P,f) \le F(b) - F(a) \le U(P,f)$$

and we must have

$$\underline{\int_a^b} f \le F(b) - F(a) \le \overline{\int_a^b} f.$$

Since $f \in \Re[a,b]$,

$$\int_a^b f = \int_a^b f(x)\,dx = F(b) - F(a).$$

5.3E1 Example: We have shown that $\frac{d}{dx}(x^n) = nx^{n-1}$ for all $x \in \mathbb{R}$ where $n \in \mathbb{N}$. Therefore, $\frac{d}{dx}\left[\frac{x^{n+1}}{n+1}\right] = x^n$ for all $x \in \mathbb{R}$ where $n \in \mathbb{Z}$, $n \ge 0$. Thus, $F(x) = \frac{x^{n+1}}{n+1}$ is an antiderivative of $f(x) = x^n$ on $\mathbb{R}$ where $n \in \mathbb{Z}$, $n \ge 0$. Therefore, we can use the above theorem to obtain: $\int_a^b x^n\,dx = \frac{b^{n+1} - a^{n+1}}{n+1}$.

5.3T2 Theorem:

Suppose $f:[a,b] \to \mathbb{R}$ is bounded and $f \in \Re[a,b]$. Define $F:[a,b] \to \mathbb{R}$ by $F(x) = \int_a^x f(t)\,dt$, for all $x \in [a,b]$. Then F is continuous, and if f is continuous at $c \in [a,b]$, then F is differentiable at c and $F'(c) = f(c)$.

Proof: We shall first show that F is uniformly continuous. Let $M > \sup_{[a,b]}|f|$ and let $x, y \in [a,b]$ with $x < y$. Then $a \le x < y \Rightarrow \int_a^y f = \int_a^x f + \int_x^y f$. Therefore,

$$\int_x^y f = \int_a^y f - \int_a^x f = F(y) - F(x).$$

Let $\varepsilon > 0$ be given and $\delta = \frac{\varepsilon}{M}$. Then

$$|F(y) - F(x)| = \left|\int_x^y f\right| \le \int_x^y |f| \le \int_x^y M = M|y - x| < \varepsilon$$

whenever $|y - x| < \delta$. Hence, F is uniformly continuous on $[a,b]$.

Suppose f is left continuous at $c \in (a,b]$ and $x \in (a,c)$. Then $F(x) - F(c) = -\int_x^c f$ and

$$\begin{aligned}\left|\frac{F(x)-F(c)}{x-c} - f(c)\right| &= \left|\frac{1}{c-x}\int_x^c f(t)\,dt - f(c)\right| \\ &= \left|\frac{1}{c-x}\int_x^c [f(t) - f(c)]\,dt\right| \\ &\le \frac{1}{c-x}\int_x^c |f(t) - f(c)|\,dt.\end{aligned}$$

Let $\varepsilon > 0$ be given. Since f is left continuous at c, there exists $\delta > 0$ such that

$$|f(t) - f(c)| < \frac{\varepsilon}{2} \text{ whenever } c - \delta < t \le c \text{ and } t \in [a,b].$$

Then for all $x \in [a,b]$ satisfying $c - \delta < x \le c$,

$$\begin{aligned}\left|\frac{F(x)-F(c)}{x-c} - f(c)\right| &\le \frac{1}{c-x}\int_x^c |f(t) - f(c)|\,dt \\ &\le \frac{1}{c-x}\int_x^c \frac{\varepsilon}{2}\,dt = \frac{\varepsilon}{2} < \varepsilon.\end{aligned}$$

Hence, $F'_-(c) = f(c)$. In a similar manner, we can show that if f is right continuous at $c \in [a,b)$, then $F'_+(c) = f(c)$. Therefore, if f is continuous at c, then f is both left continuous and right continuous at c and $F'_-(c) = f(c) = F'_+(c) = F'(c)$.

Note: If $f : [a,b] \to \mathbb{R}$ is continuous, then it always has an antiderivative on $[a,b]$ given by $F(x) = \int_a^x f(t)\,dt,\ x \in [a,b]$. Furthermore, the collection of all antiderivatives of f on $[a,b]$ is: $\{F + C \mid C \in \mathbb{R}\}$, i.e., $\left\{F(x) = \int_a^x f(t)\,dt + C,\ x \in [a,b] \,\middle|\, C \in \mathbb{R}\right\}$.

5.3E2 Example: Suppose $f:[a,b]\to\mathbb{R}$ is bounded, $f(c-)$ and $f(c+)$ exist, f is discontinuous at $c\in(a,b)$ with $f(c-)\neq f(c+)$, and $f\in\mathfrak{R}[a,b]$. Show that $F_-'(c)$ and $F_+'(c)$ exist but F is not differentiable at c.

Let $\varepsilon>0$ be given. Then there exists $\delta_1,\delta_2>0$ such that $(c-\delta_1,c+\delta_2)\subseteq[a,b]$ and

$$|f(t)-f(c-)|<\frac{\varepsilon}{2} \text{ whenever } c- \quad <t<c$$

and

$$|f(t)-f(c+)|<\frac{\varepsilon}{2} \text{ whenever } c<t<c+\delta_2.$$

If $x\in(c-\delta_1,c)$, then

$$\left|\frac{F(x)-F(c)}{x-c}-f(c-)\right|\leq\frac{1}{c-x}\int_x^c|f(t)-f(c-)|\,dt$$

$$\leq\frac{1}{c-x}\int_x^c\frac{\varepsilon}{2}\,dt=\frac{\varepsilon}{2}<\varepsilon.$$

If $x\in(c,c+\delta_2)$, then

$$\left|\frac{F(x)-F(c)}{x-c}-f(c+)\right|\leq\frac{1}{x-c}\int_c^x|f(t)-f(c+)|\,dt$$

$$\leq\frac{1}{x-c}\int_c^x\frac{\varepsilon}{2}\,dt=\frac{\varepsilon}{2}<\varepsilon.$$

Hence, $F_-'(c)=f(c-)$ and $F_+'(c)=f(c+)$. We are given that f is discontinuous at c and $f(c-)\neq f(c+)$. Thus $F'(c)$ does not exist. This example shows that integrability does not imply the existence of an antiderivative.

5.3E3 Example: The (Natural) Logarithmic Function

Let $f,L:(0,\infty)\to\mathbb{R}$ be defined by

$$f(x)=\frac{1}{x},\ x\in(0,\infty) \text{ and } L(x)=\int_1^x f=\int_1^x\frac{1}{t}\,dt,\ x\in(0,\infty).$$

Since f is continuous, $L'(x)=\dfrac{1}{x}>0$ for all $x>0$. Therefore, L is strictly increasing on $(0,\infty)$. Furthermore, from the definition of L, $L(x)<0$ for $0<x<1$, $L(1)=0$, and $L(x)>0$ for $x>1$. If we show that $\lim\limits_{x\to0+}L(x)=-\infty$ and $\lim\limits_{x\to\infty}L(x)=\infty$, then the range of L is $(-\infty,\infty)$.

(a) Suppose $a, x > 0$. Then

$$\frac{d}{dx}L(ax) = \frac{1}{ax} \cdot a = \frac{1}{x} = L'(x).$$

Therefore, $L(ax) - L(x) = C$, for all $x > 0$, where $C \in \mathbb{R}$, by (4.2T5(e)). If we let $x = 1$, then $L(a) - L(1) = C$ and since $L(1) = 0$, $L(ax) = L(a) + L(x)$. Hence,

$$L(ab) = L(a) + L(b) \text{ for all } a, b > 0.$$

(b) If we take, $b > 0$, and $a = \frac{1}{b}$, then $L(1) = L\left(\frac{1}{b}\right) + L(b)$ and

$$L\left(\frac{1}{b}\right) = -L(b),\ b > 0.$$

(c) If $n \in \mathbb{N}$ and $a > 0$, then we can use the principle of mathematical induction to prove that $L(a^n) = nL(a)$, since

$$L(a^{n+1}) = L(a^n a) = L(a^n) + L(a) = nL(a) + L(a)$$

by (a) and $L(a^1) = L(a) = 1L(a)$. Furthermore,

$$L(a^{-n}) = L\left[\left(\frac{1}{a}\right)^n\right] = nL\left(\frac{1}{a}\right) = -nL(a).$$

Therefore, $L(a^n) = nL(a),\ n \in \mathbb{Z},\ a > 0$.

If $n \in \mathbb{N}$ and $a > 0$, then $L(a) = L\left[\left(a^{1/n}\right)^n\right] = nL\left(a^{1/n}\right)$. Hence, $L\left(a^{1/n}\right) = \frac{1}{n}L(a)$ and it follows that

$$L(a^r) = rL(a),\ r \in \mathbb{Q},\ a > 0.$$

Suppose $x \in \mathbb{R}$. Then there exists a sequence $\{r_n\} \subseteq \mathbb{Q}$ such that

$$\lim_{n\to\infty} r_n = x \text{ and } \lim_{n\to\infty} a^{r_n} = a^x.$$

Since L is continuous,

$$L(a^x) = L\left(\lim_{n\to\infty} a^{r_n}\right) = \lim_{n\to\infty} L(a^{r_n}) = \lim_{n\to\infty} r_n L(a) = xL(a).$$

We have just shown that $L(a^x) = xL(a),\ x \in \mathbb{R},\ a > 0$.

(d) The number e was defined as $e = \lim_{n\to\infty}\left(1 + \frac{1}{n}\right)^n$. Since L is continuous and $L'(1) = \frac{1}{1} = 1$,

$$L(e)=L\left(\lim_{n\to\infty}\left(1+\frac{1}{n}\right)^n\right)=\lim_{n\to\infty}L\left[\left(1+\frac{1}{n}\right)^n\right]$$
$$=\lim_{n\to\infty}nL\left(1+\frac{1}{n}\right)$$
$$=\lim_{n\to\infty}\frac{L\left(1+\frac{1}{n}\right)-L(1)}{1/n}=L'(1)=1.$$

(e) Let $X\in\mathbb{R}$, $X>0$ be given. Since the harmonic series $\sum_{n=1}^{\infty}\frac{1}{n}$ diverges to ∞, so does $\sum_{n=2}^{\infty}\frac{1}{n}$. Therefore, there exists $n_0\in\mathbb{N}$ such that

$$s_{n_0}=\sum_{i=2}^{n_0}\frac{1}{i}>X.$$

Let $x\in\mathbb{R}$, $x>n_0$. Then

$$L(x)=\int_1^x\frac{1}{t}dt=\int_1^{n_0}\frac{1}{t}dt+\int_{n_0}^x\frac{1}{t}dt$$
$$>\int_1^{n_0}\frac{1}{t}dt=\sum_{i=2}^{n_0}\int_{i-1}^{i}\frac{1}{t}dt$$
$$\geq\sum_{i=2}^{n_0}\int_{i-1}^{i}\frac{1}{i}dt=\sum_{i=2}^{n_0}\frac{1}{i}>X.$$

Therefore, $\lim_{x\to\infty}L(x)=\infty.$

Since $L\left(\frac{1}{x}\right)=-L(x)$ for $x>0$,

$$\lim_{x\to0+}L(x)=\lim_{y\to\infty}L\left(\frac{1}{y}\right)=-\lim_{y\to\infty}L(y)=-\infty.$$

L is called the (natural) **logarithmic function** and is denoted by

$\log x$ (or $\log_e x$ or $\ln x$.), $x\in(0,\infty)$.

Since, L is differentiable and strictly increasing, it has an inverse, say $E:\mathbb{R}\to(0,\infty)$, which is differentiable and strictly increasing on $\mathbb{R}$. E is called the (natural) **exponential function**; it is usually denoted by

$E(x)=e^x$, $x\in\mathbb{R}$.

This function has the following properties:

(a) Let $x \in \mathbb{R}$. Then $E'(x) = \dfrac{1}{L'(E(x))} = \dfrac{1}{1/E(x)} = E(x)$.

(b) $L(1) = 0 \Rightarrow E(0) = 1$.

(c) Let $a, b \in \mathbb{R}$. If $\alpha = E(a)$ and $\beta = E(b)$, then $a = L(\alpha)$, $b = L(\beta)$. Therefore,

$$L(E(a+b)) = a + b = L(\alpha) + L(\beta)$$
$$= L(\alpha\beta) = L(E(a)E(b)).$$

It follows that $E(a+b) = E(a)E(b)$.

(d) Let $x \in \mathbb{R}$. Then

$$E(-x) = y \Rightarrow L(y) = -x$$
$$\Rightarrow x = -L(y) = L\left(\frac{1}{y}\right)$$
$$\Rightarrow E(x) = \frac{1}{y}.$$

Hence, $E(-x) = y = \dfrac{1}{E(x)}$.

(e) Let $x \in \mathbb{R}$. Then $L(e^x) = xL(e) = x$. Therefore,

$$E(x) = L^{-1}(x) = e^x,\ x \in \mathbb{R}.$$

(f) Let $a, x \in \mathbb{R}$, $a > 0$. Then

$$L(a^x) = xL(a) \Rightarrow a^x = E(xL(a)) = e^{xL(a)}.$$

5.3E4 Example: Let $f:(0,\infty) \to \mathbb{R}$ be defined by $f(x) = x^p$, $x > 0$ where $p \in \mathbb{R}$. Find the derivative and an antiderivative of *f*.

Write $f = E \circ (L \circ f)$ where $E(x) = e^x$ and $L(x) = \log x$ as defined in (5.3E3). Then

$$f(x) = E(L(x^p)) = E(pL(x))$$

and

$$f'(x) = E'(pL(x))\,pL'(x)$$
$$= E(L(x^p))\,p\frac{1}{x}$$
$$= x^p \frac{p}{x} = px^{p-1}$$

for all $x > 0$.

If $F(x) = \frac{x^{p+1}}{p+1}$, $x > 0$, $p \in \mathbb{R} \setminus \{-1\}$, then

$$F'(x) = \frac{(p+1)x^p}{p+1} = f(x)$$

and F is an antiderivative of f.

If $p = -1$, then $L(x) = \log x$ is an antiderivative of f, because $\frac{d}{dx}\log x = \frac{1}{x}$.

Note: We have now shown that the function $f(x) = x^a$, $x > 0$, $a \in \mathbb{R}$ is differentiable and its derivative is:

$$f'(x) = ax^{a-1},\ x > 0,\ a \in \mathbb{R}.$$

5.3T3 Theorem: Integration by Parts

Suppose $f, g : [a,b] \to \mathbb{R}$ are differentiable on $[a,b]$ and $f', g' \in \mathfrak{R}[a,b]$. Then

$$\int_a^b fg' = f(b)g(b) - f(a)g(a) - \int_a^b gf' = f(x)g(x)\Big|_a^b - \int_a^b gf'.$$

Proof: Since f and g are differentiable on $[a,b]$, they are continuous and thus integrable on $[a,b]$. Therefore by (5.2T1(c)), $fg', gf' \in \mathfrak{R}[a,b]$. Since $(fg)' = f'g + fg'$, $(fg)' \in \mathfrak{R}[a,b]$. Therefore, by the Fundamental Theorem of Calculus,

$$f(x)g(x)\Big|_a^b = \int_a^b (fg)' = \int_a^b f'g + \int_a^b fg'.$$

Hence result.

5.3T4 Theorem: Change of Variable

Suppose $f : [a,b] \to \mathbb{R}$ is differentiable on $[a,b]$, $f' \in \mathfrak{R}[a,b]$, and $g : I \to \mathbb{R}$ is continuous where $I = f([a,b])$. Then

$$\int_a^b g(f(t))f'(t)\,dt = \int_{f(a)}^{f(b)} g(x)\,dx.$$

Proof: Since f is differentiable, it is also continuous and $I = f([a,b])$ is a closed and bounded interval. Furthermore, $g \circ f$ is continuous. Thus $(g \circ f)f' \in \mathfrak{R}[a,b]$, by (5.2T1(c)). If I is a single point, then f is constant on $[a,b]$, and $f'(t) = 0$ for all $t \in [a,b]$ and both integrals above are zero and the result holds. Suppose I is a nontrivial interval. Let

$$G(x) = \int_{f(a)}^{x} g(s)\,ds,\ x \in I.$$

Since g is continuous, $G'(x) = g(x)$ for all $x \in I$ by (5.3T2). Furthermore, by the chain rule,

$$(G \circ f)' = (G' \circ f)f' = (g \circ f)f' \text{ on } [a,b].$$

Therefore,

$$\int_a^b g(f(t))f'(t)\,dt = (G\circ f)(t)\Big|_a^b$$

by the Fundamental Theorem of Calculus. But

$$G(f(b)) - G(f(a)) = \int_{f(a)}^{f(b)} g(x)\,dx.$$

Hence result.

5.3E5 Example: Suppose $f:[a,b]\to\mathbb{R}$ is continuous and $g,h:[c,d]\to[a,b]$ are differentiable on $[c,d]$. Define $F:[c,d]\to\mathbb{R}$ by $F(x)=\int_{g(x)}^{h(x)} f(t)\,dt,\ x\in[c,d]$. Find F'.

$$F(x)=\int_{g(x)}^{c} f(t)\,dt + \int_c^{h(x)} f(t)\,dt$$

where c is between $g(x)$ and $h(x)$. Use the chain rule:

$$\frac{d}{dx}\left[\int_{g(x)}^{c} f(t)\,dt\right] = \frac{d}{dx}\left[-\int_c^{g(x)} f(t)\,dt\right] = -f(g(x))g'(x)$$

and

$$\frac{d}{dx}\left[\int_c^{h(x)} f(t)\,dt\right] = f(h(x))h'(x).$$

Therefore, $F'(x)=f(h(x))h'(x)-f(g(x))g'(x),\ x\in[c,d]$.

5.3A1 Assignment:

(a) Let $f:\mathbb{R}\to\mathbb{R}$ be continuous and $a>0$. Define $g:\mathbb{R}\to\mathbb{R}$ by $g(x)=\int_{x-a}^{x+a} f(t)\,dt$. Show that g is differentiable and find its derivative.

(b) Let $g:\mathbb{R}\to\mathbb{R}$ be defined by $g(x)=\int_{-x^2}^{x^2} f(3t-1)\,dt$, where $f:\mathbb{R}\to\mathbb{R}$ is continuous. Find the derivative of g.

(c) Suppose $f:[0,1]\to\mathbb{R}$ is continuous. Prove that $\lim_{n\to\infty}\int_0^1 f(x^n)\,dx = f(0)$.

5.4 Improper Riemann Integrals

We will discuss the convergence of some types of improper Riemann integrals in this section and also establish the integral test for a certain class of series with nonnegative terms.

5.4D1 Definition: Improper Integrals of Type I

(a) Let $f:[a,\infty)\to\mathbb{R}$ be bounded, where $a\in\mathbb{R}$, and $f\in\Re[a,c]$ for every $c>a$. If $\lim_{c\to\infty}\int_a^c f = L\in\mathbb{R}$, then we say that the **improper Riemann integral** $\int_a^\infty f$ **converges** to L and write

$$\int_a^\infty f = L.$$

If $\lim_{c\to\infty}\int_a^c f$ does not exist, then we say that $\int_a^\infty f$ **diverges**.

If $\lim_{c\to\infty}\int_a^c f=\infty$ (or $-\infty$), then we say that $\int_a^\infty f$ diverges to ∞ (or $-\infty$) and write

$$\int_a^\infty f=\infty \text{ (or } -\infty\text{)}.$$

(b) Let $f:(-\infty,b]\to\mathbb{R}$ be bounded, where $b\in\mathbb{R}$, and $f\in\mathfrak{R}[c,b]$ for every $c<b$. If $\lim_{c\to-\infty}\int_c^b f=L\in\mathbb{R}$, then we say that the **improper Riemann integral** $\int_{-\infty}^b f$ **converges** to L and write

$$\int_{-\infty}^b f=L.$$

If $\lim_{c\to-\infty}\int_c^b f$ does not exist, then we say that $\int_{-\infty}^b f$ **diverges**.

If $\lim_{c\to-\infty}\int_c^b f=\infty$ (or $-\infty$), then we say that $\int_{-\infty}^b f$ diverges to ∞ (or $-\infty$) and write

$$\int_{-\infty}^b f=\infty \text{ (or } -\infty\text{)}.$$

(c) Let $f:\mathbb{R}\to\mathbb{R}$ be bounded. The **improper Riemann integral** $\int_{-\infty}^\infty f$ is said to **converge** if and only if both $\int_{-\infty}^a f$ and $\int_a^\infty f$ converge for some $a\in\mathbb{R}$. In this case, we write

$$\int_{-\infty}^\infty f=L+M,$$

where

$$L=\int_{-\infty}^a f \text{ and } M=\int_a^\infty f.$$

If either $\int_{-\infty}^a f$ or $M=\int_a^\infty f$ diverges for some $a\in\mathbb{R}$, then $\int_{-\infty}^\infty f$ **diverges**.

5.4E1 Example: Let $f:[1,\infty)\to\mathbb{R}$ be defined by $f(x)=\dfrac{1}{x^p}$, $x\ge 1$ where $p\in\mathbb{R}$ and $p>0$. Show that $\int_1^\infty f$ converges to $\dfrac{1}{p-1}$ if and only if $p>1$.

Suppose $0<p<1$ and $c>1$. Then

$$\int_1^c f=\int_1^c x^{-p}dx=\left.\frac{x^{-p+1}}{-p+1}\right|_1^c=\frac{1}{1-p}\left[c^{1-p}-1\right].$$

Since $\lim_{c\to\infty}c^{1-p}=\infty$, $\int_1^\infty f$ diverges to ∞.

Suppose $p>1$ and $c>1$. Then

$$\int_1^c f=\frac{1}{p-1}\left[\frac{1}{1}-\frac{1}{c^{p-1}}\right].$$

Since $\lim_{c\to\infty}\frac{1}{c^{p-1}}=0$, $\int_1^\infty f$ converges to $\frac{1}{p-1}$.

Suppose $p=1$ and $c>1$. Then $\int_1^c f=\int_1^c \frac{1}{x}dx=\log c$, because $\log 1=0$. Since $\lim_{c\to\infty}\log c=\infty$, $\int_1^\infty f$ diverges to ∞.

Note: If $p\le 0$, then $f(x)=x^{-p}\ge 1$ for all $x\ge 1$ and for $c>1$, $\int_1^c f\ge\int_1^c 1dx=c-1$. Since $\lim_{c\to\infty}(c-1)=\infty$, $\int_1^\infty f$ diverges to ∞.

5.4T1 Theorem:

Let $f:[a,\infty)\to\mathbb{R}$ be bounded, where $a\in\mathbb{R}$, $\alpha\in\mathbb{R}\setminus\{0\}$, and $f\in\Re[a,c]$ for every $c>a$. Then $\int_a^\infty f$ converges if and only if $\int_a^\infty \alpha f$ converges. In this case, $\int_a^\infty \alpha f=\alpha\int_a^\infty f$.

Proof: If $\int_a^\infty f$ converges and $c>a$, then

$$\int_a^c \alpha f=\alpha\int_a^c f\Rightarrow \lim_{c\to\infty}\int_a^c \alpha f=\lim_{c\to\infty}\alpha\int_a^c f=\alpha\lim_{c\to\infty}\int_a^c f.$$

Thus $\int_a^\infty \alpha f$ converges. If $\int_a^\infty \alpha f$ converges, then so does $\int_a^\infty \alpha^{-1}(\alpha f)=\int_a^\infty f$.

5.4T2 Theorem: Divergence Test

Let $f:[a,\infty)\to\mathbb{R}$ be decreasing and nonnegative, where $a\in\mathbb{R}$. If $\int_a^\infty f$ converges, then $\lim_{x\to\infty} f(x)=0$.

Proof: Since *f* is nonnegative on $[a,\infty)$, *f* is bounded below and $\inf f([a,\infty))=m\ge 0$. Let $\varepsilon>0$ be given, Then there exists $X\ge a$ such that

$$m\le f(X)<m+\varepsilon.$$

Since *f* is decreasing,

$$m-\varepsilon<m\le f(X)\le f(x)<m+\varepsilon \text{ for all } x>X.$$

Thus $|f(x)-m|<\varepsilon$ for all $x>X\ge a$ and $\lim_{x\to\infty} f(x)=m$.

Suppose $\lim_{x\to\infty} f(x)\ne 0$. If $c>a$, then

$$\lim_{c\to\infty}\int_a^c f\ge\lim_{c\to\infty}\int_a^c mdx=\lim_{c\to\infty}\left[m(c-a)\right]=\infty.$$

Therefore,

$$\int_a^\infty f=\infty \text{ and } \int_a^\infty f \text{ diverges.}$$

This is a contradiction. Hence result.

Note: There are analogous results for the cases where

(a) $f:[a,\infty)\to\mathbb{R}$ is nonpositive and increasing

(b) $f:(-\infty,b]\to\mathbb{R}$ is nonnegative and increasing

(c) $f:(-\infty,b]\to\mathbb{R}$ is nonpositive and decreasing.

5.4T3: Theorem:

Let $f:[a,\infty)\to\mathbb{R}$ be bounded and nonnegative, where $a\in\mathbb{R}$. Then either $\int_a^\infty f$ converges or diverges to ∞.

Proof: If $F(x)=\int_a^x f$ then $\lim_{x\to\infty}F(x)=\int_a^\infty f$. Furthermore,

$$F(x_2)=\int_a^{x_2} f=\int_a^{x_1} f+\int_{x_1}^{x_2} f$$

$$\geq\int_a^{x_1} f=F(x_1)$$

if $x_1<x_2$, because f is nonnegative. Thus F is increasing on $[a,\infty)$. Suppose $\sup F([a,\infty))=M\in\mathbb{R}$. Let $\varepsilon>0$ be given, Then there exists $X\geq a$ such that

$$M-\varepsilon<F(X)\leq M.$$

Since F is increasing,

$$M-\varepsilon<F(X)\leq F(x)\leq M<M+\varepsilon \text{ for all } x>X.$$

Therefore,

$$|F(x)-M|< \quad \text{for all } x>X\geq a$$

and

$$\lim_{x\to\infty}F(x)=M.$$

Suppose $\sup F([a,\infty))\neq M\in\mathbb{R}$. Then $\sup F([a,\infty))=\infty$ because $F(x)\geq 0$ for all $x\geq a$. If $K>0$ is given, then there exists $X\geq a$ such that $F(X)>K$. Since F is increasing,

$$F(x)\geq F(X)>K \text{ for al } x>X.$$

Hence, $\lim_{x\to\infty}F(x)=\infty$.

5.4T4 Theorem: Comparison Test

Let $f,g:[a,\infty)\to\mathbb{R}$ be bounded, where $a\in\mathbb{R}$, and $0\leq f(x)\leq g(x)$ for all $x\geq b$ for some $b\geq a$. Suppose $f,g\in\Re[a,c]$ for every $c>a$.

(a) If $\int_a^\infty g$ converges, then $\int_a^\infty f$ converges and $\int_a^\infty f\leq\int_a^\infty g$.

(b) If $\int_a^\infty f$ diverges, then $\int_a^\infty g$ diverges. In this case, $\int_a^\infty f=\int_a^\infty g=\infty$.

Proof: Since $\int_a^\infty f = \int_a^b f + \int_b^\infty f$, $\int_a^\infty f \in \mathbb{R}$ if and only if $\int_b^\infty f \in \mathbb{R}$. Therefore, it suffices to consider only the case, where $b = a$.

(a) Suppose $\int_a^\infty g$ converges. Then for every $c > a$,

$$0 \le \int_a^c f \le \int_a^c g \le \int_a^\infty g \in \mathbb{R}.$$

Therefore, $\int_a^\infty f$ converges by the previous theorem. Furthermore,

$$\int_a^c f \le \int_a^\infty g \Rightarrow \int_a^\infty f \le \int_a^\infty g.$$

(b) This is the contrapositive of the statement in (a). Since $\int_a^\infty f \le \int_a^\infty g$ and $\int_a^\infty f = \infty$, $\int_a^\infty g = \infty$.

5.4T5 Theorem: Limit Comparison Test

Let $f, g : [a, \infty) \to \mathbb{R}$ be bounded and positive, where $a \in \mathbb{R}$. Suppose

$$\lim_{x\to\infty} \frac{f(x)}{g(x)} = L > 0.$$

Then either both integrals $\int_a^\infty f$ and $\int_a^\infty g$ converge or both diverge.

Proof: There exists $b \ge a$ such that

$$\left| \frac{f(x)}{g(x)} - L \right| < \frac{L}{2} \text{ for all } x > b.$$

Therefore, for all $x > b$,

$$\frac{L}{2} g(x) < f(x) < \frac{3L}{2} g(x).$$

If $\int_a^\infty f$ converges, then so does $\int_a^\infty \frac{L}{2} g$ by (5.4T4). Therefore, $\int_a^\infty g$ converges by (5.4T1). If $\int_a^\infty g$ converges, then so does $\int_a^\infty \frac{3L}{2} g$ by (5.4T1). Hence, by (5.4T4), $\int_a^\infty f$ converges.

5.4E2 Example: Discuss the convergence of $\int_2^\infty f$, where

$$f(x) = \frac{\log x}{x^p}; x \ge 2,\ p > 0.$$

Suppose $0 < p \le 1$. Then

$$f(x) = \frac{\log x}{x^p} \ge \frac{1}{x^p} \text{ for all } x \ge e.$$

$\int_1^{\infty} \frac{1}{x^p} dx$ diverges (5.4E1). Therefore, $\int_e^{\infty} \frac{1}{x^p} dx$ diverges. Hence by (5.4T4), $\int_2^{\infty} \frac{\log x}{x^p} dx$ diverges.

Suppose $p > 1$ and let $q = \frac{p-1}{2} > 0$. Then

$$\lim_{x\to\infty} \frac{\log x}{x^q} = \lim_{x\to\infty} \frac{1/x}{qx^{q-1}} = \lim_{x\to\infty} \frac{1}{qx^q} = 0,$$

because $\lim_{x\to\infty} (\log x) = \infty = \lim_{x\to\infty} x^q$ and L'Hospital's Rule (4.2T10). Since the above limit is 0, there exists $b \geq 2$ such that for all $x > b$, $\left| \frac{\log x}{x^q} - 0 \right| < 1$. Thus, $\log x < x^q$ for all $x > b$ and

$$f(x) = \frac{\log x}{x^p} < \frac{x^q}{x^p} = \frac{1}{x^{p-q}} \text{ for all } x > b.$$

Since $p - q = p - \frac{p-1}{2} = \frac{p+1}{2} > 1$, by (5.4E1), $\int_2^{\infty} \frac{1}{x^{p-q}} dx$ converges. Therefore, $\int_2^{\infty} \frac{\log x}{x^p} dx$ converges by (5.4T4).

5.4E3 Example: Discuss the convergence of:

(a) $\int_1^{\infty} \frac{2x^2+3}{x^3+x+1} dx$ (b) $\int_1^{\infty} \frac{2x+3}{x^3+x+1} dx.$

(a) If $f(x) = \frac{2x^2+3}{x^3+x+1}$ and $g(x) = \frac{1}{x}$ for all $x \geq 1$, then $f(x), g(x) > 0$ for all $x \geq 1$ and

$$\lim_{x\to\infty} \frac{f(x)}{g(x)} = \lim_{x\to\infty} \frac{2x^3+3x}{x^3+x+1} = \lim_{x\to\infty} \frac{2+3/x^2}{6+1/x^2} = \frac{1}{3} > 0.$$

Since $\int_1^{\infty} g$ diverges by (5.4E1), then so does $\int_1^{\infty} f$ by (5.4T5).

(b) If $f(x) = \frac{2x+3}{x^3+x+1}$ and $g(x) = \frac{1}{x^2}$ for all $x \geq 1$, then $f(x), g(x) > 0$ for all $x \geq 1$ and

$$\lim_{x\to\infty} \frac{f(x)}{g(x)} = \lim_{x\to\infty} \frac{2x^3+3x^2}{x^3+x+1} = \lim_{x\to\infty} \frac{2+3/x}{6+1/x^2} = \frac{1}{3} > 0.$$

Since $\int_1^{\infty} g$ converges by (5.4E1), then so does $\int_1^{\infty} f$ by (5.4T5).

5.4D2 Definition: Improper Integrals of Type II

(a) Let $f:(a,b] \to \mathbb{R}$, where $a < b$ be such that $\int_c^b f$ exists for every $c \in (a,b)$. If $\lim_{c\to a+} \int_c^b f$ exists, then we say that the **improper Riemann integral** of f on $[a,b]$

converges and we write $\int_a^b f = \lim_{c\to a+}\int_c^b f$. If the above limit does not exist, then the integral is said to be **divergent**.

(b) Let $f:[a,b)\to\mathbb{R}$, where $a<b$ be such that $\int_a^c f$ exists for every $c\in(a,b)$. If $\lim_{c\to b-}\int_a^c f$ exists, then we say that the **improper Riemann integral** of f on $[a,b]$ **converges** and we write $\int_a^b f = \lim_{c\to b-}\int_a^c f$. If the above limit does not exist, then the integral is said to be **divergent**.

(c) Let $f:(a,b)\to\mathbb{R}$, where $a<b$ be such that $\int_c^d f$ exists for every $c,d\in(a,b)$ with $c<d$. Then the **improper Riemann integral** of f on $[a,b]$ **converges** if and only if both $\int_a^c f$ and $\int_c^b f$ converge for some $c\in(a,b)$. In this case, $\int_a^b f = \int_a^c f + \int_c^b f$.

(d) Let $f:[a,b]\backslash\{c\}\to\mathbb{R}$, where $a<c<b$ be such that $\int_a^{d_1} f$ and $\int_{d_2}^b f$ exist for $a<d_1<d_2<b$. Then the **improper Riemann integral** of f on $[a,b]$ **converges** if and only if both $\int_a^c f$ and $\int_c^b f$ converge. In this case, $\int_a^b f = \int_a^c f + \int_c^b f$.

5.4T6 Theorem:

Let $f:[a,b)\to\mathbb{R}$, where $a<b$, be nonnegative and continuous. Then the improper integral $\int_a^b f$ either converges or diverges to ∞.

Proof: The proof is quite similar to that of (5.4T3) and is left as an exercise.

5.4T7 Theorem: Comparison Test

Let $f,g:[a,b)\to\mathbb{R}$ be continuous and $0\le f(x)\le g(x)$ for all $x\in[a,b)$, where $a<b$. Then the following hold:

(a) If $\int_a^b g$ converges, then $\int_a^b f$ converges and $\int_a^b f \le \int_a^b g$.

(b) If $\int_a^b f$ diverges, then $\int_a^b g$ diverges. In this case, $\int_a^b f = \int_a^b g = \infty$.

Proof: The proof is quite similar to that of (5.4T4) and is left as an exercise.

5.4T8 Theorem: Limit Comparison Test

Let $f, g : [a,b) \to \mathbb{R}$, where $a < b$, be continuous, and positive for all $x \in [a,b)$. Suppose $\lim_{x \to b-} \frac{f(x)}{g(x)} = L > 0$. Then either both integrals $\int_a^b f$ and $\int_a^b g$ converge or both diverge.

Proof: The proof is quite similar to that of (5.4T5) and is left as an exercise.

Note: Theorems (5.4T6) – (5.4T8) have analogous versions for the improper integrals defined in (5.4D2 (a), (c), (d)).

5.4E4 Example: Let $f : (0,1] \to \mathbb{R}$ be defined by $f(x) = \frac{1}{x^p}$, $0 < x \le 1$ where $p \in \mathbb{R}$ and $p > 0$. Show that $\int_0^1 f$ converges to $\frac{1}{1-p}$ if and only if $p < 1$.

Suppose $0 < p < 1$ and $0 < c \le 1$. Then

$$\int_c^1 f = \int_c^1 x^{-p} dx = \frac{1}{1-p}\left[1 - c^{1-p}\right].$$

Therefore,

$$\int_0^1 f = \lim_{c \to 0+} \int_c^1 f = \frac{1}{1-p}[1-0] = \frac{1}{1-p}.$$

Suppose $p > 1$ and $0 < c \le 1$. Then

$$\int_c^1 f = \frac{1}{-p+1}\left[\frac{1}{1} - \frac{1}{c^{p-1}}\right].$$

Since $\lim_{c \to 0+} \frac{1}{c^{p-1}} = \infty$, $\int_1^\infty f$ diverges to ∞.

If $p = 1$ and $0 < c \le 1$, then

$$\int_c^1 f = \int_c^1 \frac{1}{x} dx = -\log c$$

because $\log 1 = 0$. Since $\lim_{c \to 0+} \log c = -\infty$, $\int_0^1 f$ diverges to ∞.

If $p \le 0$, then $f(x) = x^{-p}$ is continuous on $[0,1]$ and $\int_0^1 f$ is simply a Riemann integral and $\int_0^1 f = \frac{1}{1-p}$.

5.4A1 Assignment: Find the range of values of p for which $\int_0^e x^p \log x dx$ converges.

5.4T9 Theorem: Integral Test

Let $f:[1,\infty)\to\mathbb{R}$ be continuous, nonnegative and decreasing on $[1,\infty)$, and define $a_n = f(n)$, $n\in\mathbb{N}$. Then $\sum_{n=1}^{\infty} a_n$ converges if and only if $\int_1^{\infty} f$ converges.

Proof: Suppose $\sum_{n=1}^{\infty} a_n$ converges to s and let $t>1$. Then there exists $n\in\mathbb{N}$ such that $n\le t<n+1$. Since f is decreasing,

$$\int_1^{n+1} f = \sum_{i=1}^{n}\int_i^{i+1} f \le \sum_{i=1}^{n}\int_i^{i+1} a_i di = \sum_{i=1}^{n} a_i = s_n.$$

Therefore,

$$\int_1^{n} f \le \int_1^{t} f \le \int_1^{n+1} f \le s_n \le s$$

because $\{s_n\}$ is increasing and bounded above by s. Since $t>1$ was arbitrary, $\left\{\int_1^t f \,\middle|\, t>1\right\}$ is bounded above by s and $\int_1^{\infty} f$ converges by (5.4T3). Suppose $\int_1^{\infty} f$ converges to L and let $n\in\mathbb{N}$. Since f is nonnegative, $\int_1^{\infty} f$ is bounded above by L and

$$\begin{aligned} s_n = a_1 + \sum_{i=1}^{n-1} a_{i+1} &\le a_1 + \sum_{i=1}^{n-1}\int_i^{i+1} a_{i+1} di \\ &\le a_1 + \sum_{i=1}^{n-1}\int_i^{i+1} f = a_1 + \int_1^{n} f \le a_1 + L. \end{aligned}$$

Hence $\sum_{n=1}^{\infty} a_n$ converges because $\{s_n\}$ is increasing and bounded above.

Note: In the above theorem, the interval $[1,\infty)$ can be replaced by $[m,\infty)$ where m is an integer different from 1 along with $\sum_{n=1}^{\infty} a_n$ being replaced by $\sum_{n=m}^{\infty} a_n$. Furthermore, the condition f is decreasing can be replaced by f is decreasing on $[m,\infty)$.

5.4E5 Example: Find the range of values of p for which the series $\sum_{n=2}^{\infty}\frac{\log n}{n^p}$ converges.

Let $f(x)=\frac{\log x}{x^p}$, $x\ge 2$. Then $f(n)=\frac{\log n}{n^p}=a_n$ for all $n\ge 2$. f is clearly nonnegative on $[2,\infty)$. Furthermore,

$$f'(x)=\frac{(1/x)x^p-(\log x)px^{p-1}}{x^{2p}}=\frac{1-p\log x}{x^{p+1}}<0$$

for all $x > e^{1/p}$. Therefore, f is decreasing on $[m, \infty)$ where $m = \left[e^{1/n}\right] + 1$. We have already shown that $\int_2^{\infty} \frac{\log x}{x^p} dx$ converges if and only if $p > 1$. Hence, by the integral test, $\sum_{n=2}^{\infty} \frac{\log n}{n^p}$ converges if and only if $p > 1$.

5.4A2 Assignment: Find the values of p for which $\sum_{n=2}^{\infty} \frac{1}{n^p \log n}$ and $\sum_{n=2}^{\infty} \frac{1}{n(\log n)^p}$ converge.

CHAPTER

Sequences and Series of Functions

6.1 Sequence of Functions

In this section, we shall define a sequence of functions and derive conditions for interchange of limits to be possible. Furthermore, conditions for differentiability and integrability of the limiting function will also be discussed.

6.1D1 Definition: Sequence of Functions and Pointwise Convergence

For each $n \in \mathbb{N}$, let $f_n : E \to \mathbb{R}$, where $E \subseteq \mathbb{R}$. $\{f_n\}$ is called a **sequence of functions** on *E*. Suppose for every $x \in E$, the sequence $\{f_n(x)\}$ converges. Then we can define a function $f : E \to \mathbb{R}$ by

$$f(x) = \lim_{n\to\infty} f_n(x), \ x \in E$$

and say that $\{f_n\}$ **converges pointwise** to *f* on *E*.

Note: $\{f_n\}$ converges pointwise to *f* on *E* if and only if for each $x \in E$, given $\varepsilon > 0$, there exists $n_0 \in \mathbb{N}$ such that

$$|f_n(x) - f(x)| < \varepsilon \text{ whenever } n \geq n_0.$$

In this case, $n_0 = n_0(\varepsilon, x)$.

6.1E1 Example: Let $E = (-1,1]$ and for each $n \in \mathbb{N}$, define $f_n(x) = x^n$, $x \in E$. Then f_n is continuous on *E*. If

$$f(x) = \begin{cases} 0, & -1 < x < 1 \\ 1, & x = 1, \end{cases}$$

then $\lim_{n\to\infty} f_n(x) = f(x)$ for all $x \in E$. (This has been shown earlier.) Therefore, $\{f_n\}$ converges pointwise to *f* on *E*. However, *f* is not continuous at 1.

6.1E2 Example: Let $\{x_n\}$ be an enumeration of the rational numbers in $[0,1]$ and for each $n \in \mathbb{N}$, define

$$f_n(x) = \begin{cases} 0, & x = x_k \text{ for } 1 \le k \le n \\ 1, & \text{otherwise.} \end{cases}$$

For each $n \in \mathbb{N}$, f_n is continuous except at the points $x_1, x_2, \dots, x_n$ and thus $f_n \in \Re[0,1]$. However,

$$f(x) = \lim_{n\to\infty} f_n(x) = \begin{cases} 0, & x \in \mathbb{Q} \cap [0,1] \\ 1, & x \notin \mathbb{Q},\ x \in [0,1] \end{cases}$$

and $f \notin \Re[0,1]$.

6.1A1 Assignment:

(a) Let $E = [0,1]$ and for $n \in \mathbb{N}$, define $f_n(x) = nx(1-x^2)^n, x \in E$.

Find $f = \lim_{n\to\infty} f_n$ and show that $f_n, f \in \Re[0,1]$ but $\lim_{n\to\infty} \int_0^1 f_n \ne \int_0^1 f$.

(b) Let $E = [0,\infty)$ and for $n \in \mathbb{N}$, define $f_n(x) = \dfrac{e^{-nx}}{n}, x \in E$.

Find $f \quad \lim_{\to\infty} f$ and show that f and f_n, $n \in \mathbb{N}$ are differentiable on E, but

$$f'(x) \ne \lim_{n\to\infty} f_n'(x),\ x \in E.$$

6.1D2 Definition: Uniform Convergence

A sequence of functions $\{f_n\}$ defined on $E \subseteq \mathbb{R}$ is said to be **uniformly convergent** to a function f on E if for every $\varepsilon > 0$, there exists $n_0 \in \mathbb{N}$ such that

$$|f_n(x) - f(x)| < \varepsilon \text{ for all } n \ge n_0 \text{ and all } x \in E.$$

Note: $\{f_n\}$ converges uniformly to f on $E \Rightarrow \{f_n\}$ converges pointwise to f on E.

6.1E3 Example: Let $E = (-1,1)$ and for each $n \in \mathbb{N}$, define

$$f_n(x) = x^n,\ x \in E.$$

Then $\{f_n\}$ converges pointwise to f on E, where

$$f(x) = 0,\ x \in (-1,1).$$

Suppose $\{f_n\}$ converges uniformly to f on E. Then there exists $n_0 \in \mathbb{N}$ such that

$$|f_n(x) - f(x)| < \frac{1}{2} \text{ for all } n \ge n_0 \text{ and all } x \in E.$$

In particular,

$$|f_{n_0}(x) - f(x)| < \frac{1}{2} \text{ for all } x \in E.$$

Then

$$1 = \sup\left\{|x|^{n_0} = \left|f_{n_0}(x) - f(x)\right| \big| |x| < 1\right\} \le \frac{1}{2}.$$

This is a contradiction. Therefore, *f* is not uniformly continuous on *E*.

6.1T1 Theorem: Cauchy Criterion

Let $\{f_n\}$ be defined on $E \subseteq \mathbb{R}$. Then $\{f_n\}$ converges uniformly on *E*, if and only if for every $\varepsilon > 0$ there exists $n_0 \in \mathbb{N}$ such that

$$|f_m(x) - f_n(x)| < \varepsilon \text{ for all } m, n \ge n_0 \text{ and for all } x \in E.$$

Proof: Suppose $\{f_n\}$ converges uniformly to *f* on *E*, where $f : E \to \mathbb{R}$. If $\varepsilon > 0$ is given, then there exists $n_0 \in \mathbb{N}$ such that

$$|f_n(x) - f(x)| < \frac{\varepsilon}{2} \text{ for all } n \ge n_0 \text{ and all } x \in E.$$

Therefore, if $m, n \ge n_0$ and $x \in E$, then

$$\begin{aligned} |f_m(x) - f_n(x)| &\le |f_m(x) - f(x)| + |f(x) - f_n(x)| \\ &< \frac{\varepsilon}{2} + \frac{\varepsilon}{2} = \varepsilon. \end{aligned}$$

Conversely, suppose the Cauchy criterion holds. Then for each $x \in E$, $\{f_n(x)\}$ is a Cauchy sequence and must converge to a point, which we denote by $f(x) \in \mathbb{R}$. Therefore,

$$f(x) = \lim_{n\to\infty} f_n(x), \; x \in E.$$

Let $\varepsilon > 0$ be given. Then there exists $n_0 \in \mathbb{N}$ such that $m, n \ge n_0$ and for all $x \in E$,

$$|f_m(x) - f_n(x)| < \varepsilon / 2.$$

Fix $n \ge n_0$ and let $m \to \infty$. Then for all $n \ge n_0$ and for all $x \in E$,

$$\begin{aligned} |f(x) - f_n(x)| &= \lim_{m\to\infty} |f_m(x) - f_n(x)| \\ &\le \varepsilon / 2 < \varepsilon. \end{aligned}$$

Thus, $\{f_n\}$ converges uniformly to *f* on *E*.

6.1T2 Theorem:

Suppose $\{f_n\}$ converges pointwise to $f : E \to \mathbb{R}$. Let

$$M_n = \sup\left\{|f_n(x) - f(x)| \big| x \in E\right\}$$

for each $n \in \mathbb{N}$. Then $\{f_n\}$ converges uniformly to *f* on *E* if and only if $\lim_{n\to\infty} M_n = 0$.

Proof: Follows immediately because

$$|f_n(x) - f(x)| \le \varepsilon \text{ for all } x \in E \text{ if and only if } M_n \le \varepsilon$$

for any arbitrary $\varepsilon > 0$.

6.1E4 Example: Let $E = [0,1]$ and for $n \in \mathbb{N}$, define

$$f_n(x) = nx(1-x^2)^n, x \in E.$$

Find $f = \lim_{n\to\infty} f_n$ and show that $\{f_n\}$ does not converge uniformly to *f* on *E*.

In (6.1A1(a)), we have established that $f = \lim_{\to\infty} f = 0$. Let $n \in \mathbb{N}$ be fixed. Then

$$f_n'(x) = n(1-x^2)^n + n^2x(1-x^2)^{n-1}\cdot(-2x)$$
$$= n(1-x^2)^{n-1}(1-(1+2n)x^2) = 0$$

if $x = 1$ or $1/\sqrt{1+2n}$ because $x \in [0,1]$. The extreme values of f_n are at $x = 0, 1,$ and $1/\sqrt{1+2n}$. These values are:

$$f_n(0) = f_n(1) = 0,$$
$$f_n\left(\frac{1}{\sqrt{1+2n}}\right) = n\frac{1}{\sqrt{1+2n}}\left(1-\frac{1}{1+2n}\right)^n = \frac{n(2n)^n}{(1+2n)^{n+1/2}}.$$

Therefore,

$$\min_E f_n = 0, \ \max_E f_n = \frac{n(2n)^n}{(1+2n)^{n+1/2}},$$

$$M_n = \sup_E |f_n - f| = \frac{n(2n)^n}{(1+2n)^{n+1/2}}.$$

It follows that $\lim_{n\to\infty} M_n = \infty \neq 0$. Hence, $\{f_n\}$ does not converge uniformly to *f* on *E*.

6.1A2 Assignment: Suppose $f : \mathbb{R} \to \mathbb{R}$ is uniformly continuous and for each $n \in \mathbb{N}$, let $f_n(x) = f\left(x + \frac{1}{n}\right)$, $x \in \mathbb{R}$. Prove that $\{f_n\}$ converges uniformly to *f* on $\mathbb{R}$.

6.1D3 Definition:

A sequence of functions $\{f_n\}$ defined on $E \subseteq \mathbb{R}$ is said to be uniformly bounded on *E* if there exists $M > 0$ such that

$$|f_n(x)| \le M \text{ for all } x \in E \text{ and } n \in \mathbb{N}.$$

6.1A3 Assignment: Suppose $\{f_n\}$ is a sequence of bounded functions on *E* that converges uniformly, where $E \subseteq \mathbb{R}$. Show that $\{f_n\}$ is uniformly bounded on *E*.

6.1A4 Assignment: Suppose $\{f_n\}$ and $\{g_n\}$ converge uniformly to *f* and g, respectively, on *E*, where $E \subseteq \mathbb{R}$. Prove the following:

(a) $\{f_n + g_n\}$ converges uniformly to $f + g$ on *E*.

(b) If $\{f_n\}$ and $\{g_n\}$ are sequences of bounded functions, then $\{f_n g_n\}$ converges uniformly to fg on *E*.

6.1T3 Theorem:

Suppose $\{f_n\}$ converges uniformly to *f* on *E*, where $E \subseteq \mathbb{R}$. Let *p* be a limit point of *E* and suppose for each $n \in \mathbb{N}$,

$$\lim_{x \to p} f_n(x) = a_n \in \mathbb{R}.$$

Then $\{a_n\}$ converges and

$$\lim_{x \to p} f(x) = \lim_{n \to \infty} a_n.$$

This can be written as

$$\lim_{x \to p}\left(\lim_{n \to \infty} f_n(x)\right) = \lim_{n \to \infty}\left(\lim_{x \to p} f_n(x)\right).$$

Proof: Let $\varepsilon > 0$ is given. Then there exists $n_0 \in \mathbb{N}$ such that

$$|f_m(x) - f_n(x)| < \varepsilon \text{ for all } m, n \geq n_0 \text{ and all } x \in E.$$

Therefore,

$$\lim_{x \to p}|f_m(x) - f_n(x)| \leq \varepsilon, \text{ i.e., } |a_m - a_n| \leq \varepsilon \text{ for all } m, n \geq n_0.$$

Thus $\{a_n\}$ is a Cauchy sequence and converges, say to *a*. Again let $\varepsilon > 0$ is given. Then there exists $m \in \mathbb{N}$ such that

$$|f(x) - f_m(x)| < \frac{\varepsilon}{3} \text{ for all } x \in E \text{ and } |a_m - a| < \frac{\varepsilon}{3}.$$

Since $\lim_{x \to p} f_n(x) = a_n$, there exists $\delta > 0$ such that

$$|f_m(x) - a_m| < \frac{\varepsilon}{3} \text{ for all } x \in E \text{ with } 0 < |x - p| < \delta.$$

Then

$$\begin{aligned} |f(x) - a| &\leq |f(x) - f_m(x)| + |f_m(x) - a_m| + |a_m - a| \\ &\leq \frac{\varepsilon}{3} + \frac{\varepsilon}{3} + \frac{\varepsilon}{3} = \varepsilon \end{aligned}$$

whenever $0 < |x - p| < \delta$. Hence result.

Note: If *f* is continuous at *p*, then $\lim_{x \to p}\left(\lim_{n \to \infty} f_n(x)\right) \quad \lim_{n \to \infty} f_n(p)$.

6.1T3C Corollary:

For each $n \in \mathbb{N}$, let $f_n : E \to \mathbb{R}$ be continuous and suppose $\{f_n\}$ converges uniformly to *f* on *E*, where $E \subseteq \mathbb{R}$. Then *f* is continuous on *E*.

Proof: If $p \in E$ is an isolated point of *E*, then *f* is continuous at *p*. Suppose *p* is a limit point of *E*. Since f_n is continuous at *p* for each $n \in \mathbb{N}$,

$$\lim_{x \to p} f_n(x) = f_n(p).$$

Therefore, by the above theorem,

$$\lim_{x \to p} f(x) = \lim_{x \to p} \lim_{n \to \infty} f_n(x) = \lim_{n \to \infty} \lim_{x \to p} f_n(x) = \lim_{n \to \infty} f_n(p) = f(p).$$

Hence result.

6.1E5 Example: Let $E = [0,1]$ and for each $n \in \mathbb{N}$, define $f_n(x) = x^n$, $x \in E$. Show that $\lim_{n \to \infty} f_n$ is not uniformly continuous on E.

For all $\in \mathbb{N}$ is continuous on *E*. If

$$f(x) = \begin{cases} 0, & 0 \le x < 1 \\ 1, & x = 1, \end{cases}$$

then $\lim_{n \to \infty} f_n(x) = f(x)$ for all $x \in E$. (This has been shown earlier.) But *f* is not continuous. Therefore, *f* cannot be uniformly continuous on $[0,1]$.

Note: From the previous corollary, we see that if a sequence of continuous functions converges uniformly, then the limiting function is continuous. However, the converse of this is not true as can be seen by the next example.

6.1E6 Example: Let

$$f_n(x) = nxe^{-nx^2}, x \in [0,1] \text{ and } n \in \mathbb{N}.$$

Then $\{f_n\}$ is a sequence of continuous functions that converges to the continuous function $f:[0,1] \to \mathbb{R}$ defined by $f(x) = 0$, $x \in [0,1]$. We can use L'Hospital's rule to show:

$$f_n'(x) = ne^{-nx^2} - 2n^2x^2e^{-nx^2} = 0$$

if $x = 1/\sqrt{2n}$ because $x \in [0,1]$. *f* is clearly continuous. Since

$$f_n(0) = 0, f_n(1) = ne^{-n},$$

$$f_n\left(1/\sqrt{2n}\right) = \sqrt{n/2}e^{-1/2} = \sqrt{n/2e}$$

are the only possible extreme values of f_n,

$$\max_{[0,1]} f_n = \sqrt{n/2e} = M_n \text{ for } n \in \mathbb{N}.$$

Since $\lim_{n \to \infty} M_n = \infty \ne 0$, $\{f_n\}$ cannot be uniformly continuous.

6.1A5 Assignment: Let

$$f_n(x) = \frac{x^n}{1+x^n}, \ x \in [0,1], \text{ for all } n \in \mathbb{N}.$$

(a) Show that $\{f_n\}$ converges uniformly to *f* on $[0,a]$ where $a \in (0,1)$ and $f(x) = 0$ for all $x \in \mathbb{R}$.

(b) Does *f* converge uniformly on $[0,1]$? Justify your answer!

6.1T4 Theorem: Dini's Theorem

Let *K* be a compact subset of $\mathbb{R}$ and let $f, f_n : K \to \mathbb{R}$ be continuous for $n \in \mathbb{N}$. Suppose

(a) $\{f_n\}$ converges pointwise to *f* on *K*, and

(b) $f_n(x) \geq f_{n+1}(x)$ for all $x \in K$ and $n \in \mathbb{N}$.

Then $\{f_n\}$ converges uniformly to *f* on *K*.

Proof: Using (a) and (b), we see that

$$f_n(x) \geq f_{n+1}(x) \geq f(x) \text{ for all } x \in K \text{ and } n \in \mathbb{N}.$$

If $g_n = f_n - f$ for all $n \in \mathbb{N}$. Then g_n is continuous and

$$g_n(x) \geq g_{n+1}(x) \geq 0 \text{ for all } x \in K \text{ and } n \in \mathbb{N}.$$

Furthermore, $\{g_n\}$ converges pointwise to 0 on *K*. Let $\varepsilon > 0$ be given and for each $n \in \mathbb{N}$, let $K_n = \{x \in K | g_n(x) \geq \varepsilon\}$.

Since $K_n^c = \{x \in K | g_n(x) < \varepsilon\}$ is open, K_n is closed and must be compact for all $n \in \mathbb{N}$.

$$K_{n+1} \subseteq K_n, \ n \in \mathbb{N}$$

because $g_n(x) \geq g_{n+1}(x)$ for all $x \in K$ and $n \in \mathbb{N}$. Let $x \in K$. Then there exists $n_1 \in \mathbb{N}$ such that

$$|g_n(x) - 0| < \varepsilon \text{ for all } n \geq n$$

Thus $x \notin K$ and $x \notin \overset{\infty}{\ } K$ This shows that $\bigcap_{n=0}^{\infty} K_n = \phi$. Therefore, $K_{n_0} = \phi$ for some $n_0 \in \mathbb{N}$. Hence,

$$|g_n(x) - 0| < \varepsilon \text{ for all } n \geq n_0, \ x \in E.$$

Thus $\{g_n\}$ converges uniformly to 0 on *K* and $\{f_n\}$ converges uniformly to *f* on *K*.

6.1E7 Example: Define the sequence $\{f_n\}$ by

$$f_n(x) = \frac{1}{1+nx}, \ x \in E = (0,1) \text{ and } n \in \mathbb{N}.$$

Then $\{f_n\}$ converges pointwise to 0 on E and

$$f_n(x)=\frac{1}{1+nx}\geq\frac{1}{1+(n+1)x}=f_{n+1}(x) \text{ for all } x\in E \text{ and } n\in\mathbb{N}.$$

However,

$$M_n=\sup\{|f_n(x)-f(x)|\,|\,x\in E\}=\frac{1}{1+0}=1=\lim_{n\to\infty}M_n\neq 0.$$

Therefore by (6.1T2), $\{f_n\}$ does not converge uniformly to 0 on E.

6.1A6 Assignment: Define the sequence $\{f_n\}$ by $f_n(x)=\left(1+\frac{x}{n}\right)^n$ for all $x\in\mathbb{R}$ and $n\in\mathbb{N}$. Use Dini's theorem to prove that $\{f_n\}$ converges uniformly to the function $f:[a,b]\to\mathbb{R}$ defined by $f(x)=e^x$, $x\in[a,b]$ for any fixed $a,b\in\mathbb{R}$ with $a<b$.

6.1D4 Definition: Norm and Normed Linear Space

It is assumed that the reader is familiar with the concept of a vector space over $\mathbb{R}$. Let V be a vector space over $\mathbb{R}$. Then $\|\ \|:V\to\mathbb{R}$ is called a **norm** on V if it satisfies the following:

(a) $\|x\|\geq 0$ for all $x\in V$ with equality if and only if $x=0$.

(b) $\|cx\|=|c|\|x\|$ for all $x\in V$ and $c\in\mathbb{R}$.

(c) $\|x+y\|\leq\|x\|+\|y\|$ for all $x,y\in\mathbb{R}$.

The pair $(V,\|\ \|)$ is called a **normed linear space**.

6.1E8 Example: Define $\|\ \|:\mathbb{R}\times\mathbb{R}\to\mathbb{R}$ by

$$\|x\|=\|(x_1,x_2)\|=\max\{|x_1|,|x_2|\},\text{ for all } x=(x_1,x_2)\in\mathbb{R}\times\mathbb{R}.$$

(a) $0\leq\|x\|=\|(x_1,x_2)\|=\max\{|x_1|,|x_2|\}=0$ if and only if

$$\Rightarrow x=(x_1,x_2)=(0,0).$$

(b) $\|cx\|=\|c(x_1,x_2)\|=\max\{|c||x_1|,|c||x_2|\}=|c|\max\{|x_1|,|x_2|\}=|c|\|x\|.$

(c) $\|x+y\|=\max\{|x_1+y_1|,|x_2+y_2|\}\leq\max\{|x_1|+|y_1|,|x_2|+|y_2|\}$

$$\leq\max\{\|x\|+\|y\|,\|x\|+\|y\|\}=\|x\|+\|y\|$$

This shows that $\{\mathbb{R}\times\mathbb{R},\|\ \|\}$ is a normed linear space.

6.1D5 Definition: The Normed Linear Space $C[a,b]$

Define the space

$$C[a,b] = \{ f : [a,b] \to \mathbb{R} \mid f \text{ is continuous} \},$$

where $a, b \in \mathbb{R}$ with $a < b$. Then, with the usual definitions of addition and scalar multiplication, $C[a,b]$ is a vector space over $\mathbb{R}$. This can easily be verified. Define a norm

$$\|f\|_\infty = \max\{ |f(x)| \mid x \in [a,b] \},\ f \in C[a,b].$$

Then by (6.1A7) below, $\{C[a,b], \| \ \|_\infty\}$ is a normed linear space.

6.1A7 Assignment: Show that $\| \ \|_\infty$ is a norm on $C[a,b]$.

6.1T5 Theorem:

A sequence $\{f_n\}$ in $C[a,b]$ converges uniformly to $f \in C[a,b]$ if and only if given $\varepsilon > 0$, there exists $n_0 \in \mathbb{N}$ such that

$$\|f_n - f\|_\infty < \varepsilon \text{ whenever } n \geq n_0.$$

In this case, we say that $\{f_n\}$ converges to f in **norm**.

Proof: Suppose $\{f_n\}$ in $C[a,b]$ converges uniformly to f on $[a,b]$. Then by (6.1T3C), f is continuous and is in $C[a,b]$. If $\varepsilon > 0$ is given, there exists $n_0 \in \mathbb{N}$ such that

$$|f_n(x) - f(x)| < \varepsilon \text{ for all } n \geq n_0 \text{ and } x \in [a,b].$$

Since $|f_n - f| \in C[a,b]$,

$$\|f_n - f\|_\infty = \max\{ |f_n(x) - f(x)| \mid x \in [a,b] \}.$$

Therefore, $\|f_n - f\|_\infty < \varepsilon$ for all $n \geq n_0$.

Conversely, suppose given $\varepsilon > 0$, there exists $n_0 \in \mathbb{N}$ such that

$$\|f_n - f\|_\infty < \varepsilon \text{ whenever } n \geq n_0,$$

where $f, f_n \in C[a,b]$. Then

$$|f_n(x) - f(x)| < \varepsilon \text{ for all } n \geq n_0 \text{ and } x \in [a,b],$$

i.e., $\{f_n\}$ converges uniformly to f on $[a,b]$.

6.1D6 Definition: Cauchy Sequence and Complete Normed Linear Space

Let $Y = (V, \| \ \|)$ be a normed linear space over $\mathbb{R}$.

(a) A sequence $\{x_n\}$ in Y is said to be a **Cauchy sequence** if for every $\varepsilon > 0$, there exists $n_0 \in \mathbb{N}$ such that

$$\|x_m - x_n\| < \varepsilon \text{ whenever } m, n \geq n_0.$$

(b) The normed linear space Y is said to be **complete** if each Cauchy sequence in Y converges in norm to an element in Y.

Note: The normed linear space $(\mathbb{R}, |\ |)$ is complete.

6.1T6 Theorem:

The normed linear space $(C[a,b], \|\ \|_\infty)$ is complete.

Proof: Suppose $\{f_n\}$ in $C[a,b]$ is a Cauchy sequence and let $\varepsilon > 0$ be given. Then there exists $n_0 \in \mathbb{N}$ such that

$$\|f_m - f_n\| < \varepsilon \text{ for all } m, n \geq n_0.$$

Therefore,

$$|f_n(x) - f(x)| \leq \|f_m - f_n\| < \varepsilon \text{ for all } m, n \geq n_0 \text{ and } x \in [a,b].$$

Thus by (6.1T1), $\{f_n\}$ converges uniformly to some function f on $[a,b]$. By (6.1T3C), f is continuous and belongs to $C[a,b]$. Since the convergence is uniform, if $\varepsilon > 0$ is given, then there exists $n_1 \in \mathbb{N}$ such that

$$|f_n(x) - f(x)| < \varepsilon \text{ for all } n \geq n_1 \text{ and } x \in [a,b].$$

Thus

$$\|f_n - f\|_\infty = \max\{|f_n(x) - f(x)| \,|\, x \in [a,b]\} < \varepsilon \text{ for all } n \geq n_1.$$

Hence result.

6.1D7 Definition: Contraction Mapping

Let $(V, \|\ \|)$ be a normed linear space. A mapping $T: V \to V$ is called a **contraction mapping** if there exists $c \in \mathbb{R}$ with $0 < c < 1$ such that

$$\|T(x) - T(y)\| \leq c\|x - y\| \text{ for all } x, y \in V.$$

6.1T7 Theorem: Fixed Point

Let $(V, \|\ \|)$ be a complete normed linear space and let $T: V \to V$ be a contraction mapping. Then there exists a unique point $x \in V$ such that $T(x) = x$. The point x is called a **fixed point** of T.

Proof: Suppose there exists $c \in \mathbb{R}$ with $0 < c < 1$ such that

$$\|T(x) - T(y)\| \leq c\|x - y\| \text{ for all } x, y \in V.$$

Define $\{x_n\}$ as follows: Let $x_0 \in V$ be arbitrary and

$$x_n = T(x_{n-1}) \text{ for all } n \in \mathbb{N}.$$

Then

$$\|x_{n+1} - x_n\| = \|T(x_n) - T(x_{n-1})\|$$
$$\leq c\|x_n - x_{n-1}\|$$

for all $n > 1$. It follows that

$$\|x_{n+1} - x_n\| \leq c^n \|x_1 - x_0\| \text{ for all } n \in \mathbb{N}.$$

Thus

$$\|x_{n+p} - x_n\| \leq \|x_{n+p} - x_{n+p-1}\| + \ldots + \|x_{n+1} - x_n\|$$
$$\leq c^n \left(c^{p-1} + c^{p-2} + \ldots + c^0\right)\|x_1 - x_0\|$$

for all $n, p \in \mathbb{N}$. Since $0 < c < 1$,

$$c^{p-1} + c^{p-2} + \ldots + c^0 \leq 1/(1-c).$$

Hence,

$$\|x_{n+p} - x_n\| \leq \frac{c^n}{1-c}\|x_1 - x_0\| \text{ for all } n, p \in \mathbb{N}.$$

Since $\lim_{n\to\infty} c^n = 0$, $\{x_n\}$ is a Cauchy sequence in V. Therefore, $\{x_n\}$ converges in norm to some $x \in V$ because $(V, \|\ \|)$ is complete. Let $\varepsilon > 0$ be given and let $\varepsilon' = \varepsilon/(c+1)$. Then there exists $n_0 \in \mathbb{N}$ such that

$$\|x_n - x\| < \varepsilon \text{ for all } n \geq n_0.$$

Thus

$$\|T(x) - x\| \leq \|T(x) - T(x_{n_0})\| + \|T(x_{n_0}) - x\|$$
$$\leq c\|x - x_{n_0}\| + \|x_{n_0+1} - x\| < (c+1)\varepsilon' = \varepsilon.$$

Since $\varepsilon > 0$ was arbitrary,

$$\|T(x) - x\| = 0 \text{ and } T(x) = x.$$

Suppose $T(y) = y$ for some $y \in V$. Then

$$\|x - y\| = \|T(x) - T(y)\| \leq c\|x - y\|.$$

Since $0 < c < 1$, we must have $\|x - y\| = 0$, i.e., $x = y$. Hence, x is unique.

6.1E9 Example: Let $T : C[0,1] \to C[0,1]$ be defined by

$$T(f) = F \text{ for all } f \in C[0,1],\ F(x) = \int_0^x f,\ x \in [0,1].$$

Furthermore, define $T^2 = T \circ T$. Prove the following:

(a) $\left|T^2(f)(x)\right| \le \frac{1}{2}x^2 \|f\|_\infty$ for all $x \in [0,1]$.

(b) T^2 is a contraction mapping on $C[0,1]$ and has a fixed point in $C[0,1]$.

(a) Let $x \in [0,1]$. Then

$$\begin{aligned}\left|T^2(f)(x)\right| = \left|T(F)(x)\right| = \left|\int_0^x F\right| = \left|\int_0^x \int_0^t f(s)\,dsdt\right| &\le \int_0^x \int_0^t |f(s)|\,dsdt \\ &\le \int_0^x \int_0^t \|f\|_\infty\,dsdt \\ &= \int_0^x t\|f\|_\infty\,dt = \frac{1}{2}x^2\|f\|_\infty.\end{aligned}$$

(b)
$$\begin{aligned}T^2(f)(x) - T^2(g)(x) = \int_0^x \int_0^t f(s)\,dsdt - \int_0^x \int_0^t g(s)\,dsdt &= \int_0^x \int_0^t \left[f(s) - g(s)\right]dsdt \\ &= T^2(f-g)(x) \text{ for all } x \in [0,1].\end{aligned}$$

Therefore, by (a),

$$\begin{aligned}\left|\left[T^2(f) - T^2(g)\right](x)\right| &= \left|T^2(f-g)(x)\right| \\ &\le \frac{1}{2}x^2\|f-g\|_\infty \text{ for all } x \in [0,1],\end{aligned}$$

$$\left|\left[T^2(f) - T^2(g)\right](x)\right| \le \frac{1}{2}\|f-g\|_\infty \text{ for all } x \in [0,1]$$

and

$$\left\|T^2(f) - T^2(g)\right\|_\infty \le \frac{1}{2}\|f-g\|_\infty.$$

Hence, T^2 is a contraction mapping on $C[0,1]$ and has a unique fixed point in $C[0,1]$.

6.1A8 Assignment:

Let $T: C[0,1] \to C[0,1]$ be defined by

$T(f) = F$ for all $f \in C[0,1]$ where $F(x) = \int_0^x f$, $x \in [0,1]$.

Prove that T has a fixed point in $C[0,1]$. Hint: A fixed point f of T^2 satisfies $f'' = f$ and a fixed point g of T satisfies $g' = g$. Some fixed point of T^2 would also be a fixed point of T. Show that the only fixed point of T^2 and also of T is: $f(x) = 0$, $x \in \mathbb{R}$.

6.1T8 Theorem:

Suppose for all $n \in \mathbb{N}$, $f_n \in \mathfrak{R}[a,b]$, and $\{f_n\}$ converges uniformly to f on $[a,b]$, where $a,b \in \mathbb{R}$, $a < b$. Then $f \in \mathfrak{R}[a,b]$ and $\int_a^b f = \lim_{n\to\infty} \int_a^b f_n$.

Proof: Let

$$M_n = \sup_{x\in[a,b]} |f_n(x) - f(x)|$$

(as defined previously) for each $n \in \mathbb{N}$. Then by (6.1T2), $\lim_{n\to\infty} M_n = 0$. Furthermore,

$$f_n(x) - M_n < f(x) < f_n(x) + M_n \text{ for all } x \in [a,b].$$

Therefore,

$$\int_a^b (f_n - M_n) \le \underline{\int_a^b} f \le \overline{\int_a^b} f \le \int_a^b (f_n + M_n).$$

Hence,

$$0 \le \overline{\int_a^b} f - \underline{\int_a^b} f \le 2M_n(b-a).$$

Since $\lim_{n\to\infty} M_n = 0$, $f \in \mathfrak{R}[a,b]$. Furthermore,

$$\left|\int_a^b f_n - \int_a^b f\right| \le \int_a^b |f_n - f| \le M_n(b-a) \to 0 \text{ as } n \to \infty.$$

Hence, $\int_a^b f = \lim_{n\to\infty} \int_a^b f_n$.

6.1E10 Example: Suppose for all $n \in \mathbb{N}$, $f_n \in \mathfrak{R}[a,b]$, and $\{f_n\}$ converges uniformly to $f \in \mathfrak{R}[a,b]$. Define

$$F(x) = \int_a^x f,\ x \in [a,b]$$

and for all $n \in \mathbb{N}$,

$$F_n(x) = \int_a^x f_n,\ x \in [a,b].$$

Prove that $\{F_n\}$ converges uniformly to F on $[a,b]$.

Let $\varepsilon > 0$ be given. Then there exists $n_0 \in \mathbb{N}$ such that

$$|f_n(x) - f(x)| < \varepsilon' = \frac{\varepsilon}{2(b-a)} \text{ for all } n \ge n_0 \text{ and } x \in [a,b].$$

Thus,

$$\begin{aligned} |F_n(x) - F(x)| &= \left|\int_a^x f_n - \int_a^x f\right| \le \int_a^x |f_n - f| \\ &\le \int_a^b \varepsilon' = \varepsilon'(b-a) < \varepsilon. \end{aligned}$$

Hence result.

6.1A9 Assignment: Suppose for all $n \in \mathbb{N}$, $f_n \in \Re[a,b]$, and $\{f_n\}$ converges uniformly to $f \in \Re[a,b]$. If $g \in \Re[a,b]$, then prove that $\int_a^b fg = \lim_{n\to\infty} \int_a^b f_n g$.

6.1T9 Theorem:

Let $\{f_n\}$ be a sequence of differentiable functions on $[a,b]$ and suppose $\{f_n(x_0)\}$ converges for some $x_0 \in [a,b]$. If $\{f_n'\}$ converges uniformly on $[a,b]$, then $\{f_n\}$ converges uniformly to some function *f* on $[a,b]$ and

$$\lim_{n\to\infty} f_n'(x) = f'(x),\ x \in [a,b].$$

Proof: Let $\varepsilon > 0$ be given. Then there exists $n_0 \in \mathbb{N}$ such that

$$|f_m(x_0) - f_n(x_0)| < \frac{\varepsilon}{2} \text{ and } |f_m'(t) - f_n'(t)| < \frac{\varepsilon}{2(b-a)}$$

for all $m, n \geq n_0$ and $t \in [a,b]$. Keep $m, n \geq n_0$ fixed and apply the MVT to $f_m - f_n$:

$$|(f_m(x) - f_n(x)) - (f_m(x_0) - f_n(x_0))| = |f_m'(t) - f_n'(t)||x - x_0| \text{ for all } x \in [a,b]$$

for some *t* between *x* and x_0. By the above inequalities, if $m, n \geq n_0$, then for all $x \in [a,b]$,

$$|f_m(x) - f_n(x)| \leq |(f_m(x) - f_n(x)) - (f_m(x_0) - f_n(x_0))| + |f_m(x_0) - f_n(x_0)|$$

$$< \frac{\varepsilon|x_0 - x|}{2(b-a)} + \frac{\varepsilon}{2} \leq \frac{\varepsilon(b-a)}{2(b-a)} + \frac{\varepsilon}{2} = \varepsilon.$$

Hence by (6.1T1), $\{f_n\}$ converges uniformly to some function *f* on $[a,b]$, i.e.,

$$f(x) = \lim_{n\to\infty} f(x) \text{ for all } x \in [a,b].$$

In order to show that *f* is differentiable and that $\lim_{n\to\infty} f_n'(x) = f'(x)$, $x \in [a,b]$, fix $p \in [a,b]$ and for $t \in [a,b]$, $t \neq p$ define

$$g(t) = \frac{f(t) - f(p)}{t - p},\ g_n(t) = \frac{f_n(t) - f_n(p)}{t - p},\ n \in \mathbb{N}.$$

Then for all $t \in [a,b] \backslash \{p\}$,

$$\lim_{n\to\infty} g(t) = g(t) \text{ and } \lim_{t\to p} g_n(t) = f_n'(p) \text{ for each } n \in \mathbb{N}.$$

Let $t \neq p$. Use the MVT to obtain,

$$|g_m(t) - g_n(t)| = \frac{|(f_m(t) - f_n(t)) - (f_m(p) - f_n(p))|}{|t - p|} = |f_m'(s) - f_n'(s)|$$

for some *s* between *p* and *t*. Then

$$|g_m(t) - g_n(t)| \le \frac{\varepsilon}{2(b-a)} \text{ for all } m, n \ge n_0.$$

Hence, $\{g_n\}$ converges uniformly to *g* on $[a,b]\setminus\{p\}$ and by (6.1T3),

$$f'(p) = \lim_{t\to p} g(t) = \lim_{n\to\infty} f_n'(p).$$

6.1E11 Example: Give a shorter proof for the above theorem with the added hypothesis: f_n' is continuous on $[a,b]$.

Since f_n' is continuous, by (5.3T2),

$$f_n(x) = f_n(x_0) + \int_{x_0}^{x} f_n' \text{ for all } x \in [a,b].$$

Let $\varepsilon > 0$ be given. Then there exists $n_0 \in \mathbb{N}$ such that

$$\left|f_m'(t) - f_n'(t)\right| < \frac{\varepsilon}{2(b-a)} \text{ for all } m, n \ge n_0 \text{ and } t \in [a,b].$$

Thus

$$\begin{aligned} |f_m(x) - f_n(x)| &\le \int_{x_0}^{x} \left|f_m'(t) - f_n'(t)\right| dt \\ &\le \frac{\varepsilon}{2(b-a)} |x - x_0| < \varepsilon \end{aligned}$$

for all $x \in [a,b]$. Hence, $\{f_n\}$ converges uniformly to some function *f* on $[a,b]$. Let

$$\lim_{n\to\infty} f_n'(x) = g(x) \text{ for all } x \in [a,b].$$

Then

$$\begin{aligned} f(x) &= \lim_{n\to\infty} f_n(x) \\ &= \lim_{n\to\infty}\left[f_n(x_0) + \int_{x_0}^{x} f_n'\right] \\ &= f(x_0) + \int_{x_0}^{x} \lim_{n\to\infty} f_n' = f(x_0) + \int_{x_0}^{x} g \end{aligned}$$

for all $x \in [a,b]$ by (6.1T8), and

$$f'(x) = g(x) = \lim_{n\to\infty} f_n'(x)$$

for all $x \in [a,b]$.

6.1A10 Assignment: For each $n \in \mathbb{N}$, let $f_n : (a,b) \to \mathbb{R}$ be differentiable and suppose $\{f_n\}$ converges pointwise to a function f on (a,b). If $\{f_n\}$ converges uniformly on each compact subset of (a,b), then prove that *f* is differentiable on (a,b) and

$$f'(x) = \lim_{n\to\infty} f_n'(x) \text{ for all } x \in (a,b).$$

Hint: If $p \in (a,b)$, then there exists $c, d \in (a,b)$ such that $a < c < p < d < b$.

6.2 Series of Functions

In this section, we shall define a series of functions and derive conditions for interchange of limits to be possible. Furthermore, conditions for differentiability and integrability of the limiting function will also be discussed.

6.2D1 Definition: Series of Functions and Pointwise Convergence

For each $n \in \mathbb{N}$, let $f_n : E \to \mathbb{R}$, where $E \subseteq \mathbb{R}$ and let

$$s_n(x) = \sum_{k=1}^{n} f_k(x).$$

The sequence of **partial sums** $\{s_n\}$ is called a series of functions on *E*, and is denoted by $\sum_{n=1}^{\infty} f_n$. The series $\sum_{n=1}^{\infty} f_n$ is said to be **pointwise convergent** on *E*, if for each $x \in E$, the series $\sum_{n=1}^{\infty} f_n$ converges. If the series converges pointwise to *s* on *E*, then we write

$$s = \sum_{n=1}^{\infty} f_n \text{ or } s(x) = \sum_{n=1}^{\infty} f_n(x),\ x \in E;$$

s is called the **sum** of the series. We can also define a series of functions

$$\sum_{n=m}^{\infty} f_n,\ m \in \mathbb{Z},$$

with suitable modifications to the above definition.

6.2E1 Example: Let $a \in \mathbb{R}$, $a \neq 0$. Determine the set of all points at which the series $\sum_{n=1}^{\infty} ax^{n-1}$ converges pointwise and find its sum.

$$s_n(x) = \sum_{n=1}^{\infty} ax^{n-1} = \begin{cases} a\dfrac{1-x^n}{1-x}, & x \neq 1 \\ na, & x = 1. \end{cases}$$

It follows that $\{s_n(x)\}$ converges to $s(x) = \dfrac{a}{1-x}$ if and only if $|x| < 1$. Hence, $\{s_n\}$ converges pointwise to *s* on $(-1,1)$ and

$$\sum_{n=1}^{\infty} ax^{n-1} = \frac{a}{1-x},\ x \in (-1,1).$$

6.2D2 Definition: Uniform Convergence

The series of functions $\sum_{n=1}^{\infty} f_n$ is said to be **uniformly convergent** on *E*, if the sequence of partial sums $\{s_n\}$ is uniformly convergent on *E*, where $E \subseteq \mathbb{R}$.

6.2E2 Example: The series $\sum_{n=1}^{\infty} ax^{n-1}$ converges pointwise to $s(x) = \dfrac{a}{1-x}$ on $(-1,1)$. Show that it is not uniformly convergent on $(-1,1)$.

From the previous example,

$$|s_n(x) - s(x)| = \left| a\frac{1-x^n}{1-x} - \frac{a}{1-x} \right| = \frac{|a||x|^n}{1-x}.$$

It follows that

$$M_n = \sup_{(-1,1)} |s_n(x) - s(x)| = \sup_{(-1,1)} \frac{|a||x|^n}{1-x}$$

$$\geq \sup_{(-1,1)} \frac{|a||x|^n}{2} = \frac{|a|}{2} > 0.$$

The result follows by (6.1T2).

6.2T1 Theorem: Cauchy Criterion

The series of functions $\sum_{n=1}^{\infty} f_n$ converges uniformly on $E \subseteq \mathbb{R}$ if and only if for every $\varepsilon > 0$, there exists $n_0 \in \mathbb{N}$ such that

$$\left| \sum_{k=n+1}^{m} f_k(x) \right| < \varepsilon \text{ for all } x \in E \text{ and } m > n \geq n_0.$$

Proof: Follows directly from (6.1T1) applied to $\{s_n\}$.

6.2T2 Theorem: Weierstrass M-Test

Suppose for each $n \in \mathbb{N}$, $f_n : E \to \mathbb{R}$, where $E \subseteq \mathbb{R}$ and

$$|f_n(x)| \leq M_n \in \mathbb{R} \text{ for all } x \in E.$$

If $\sum_{n=1}^{\infty} M_n$ converges, then $\sum_{n=1}^{\infty} f_n$ converges uniformly and absolutely on *E*.

Proof: Let $\varepsilon > 0$ be given. Then there exists $n_0 \in \mathbb{N}$ such that

$$\left| \sum_{k=n+1}^{m} M_k \right| < \varepsilon \text{ for all } m > n \geq n_0.$$

Therefore,

$$\left|\sum_{k=n+1}^{m} f_k(x)\right| \leq \sum_{k=n+1}^{m} |f_k(x)| \leq \left|\sum_{k=n+1}^{m} M_k\right| < \varepsilon \quad \text{for all } m > n \geq n_0.$$

Result follows by (6.2T1).

6.2E3 Example: Show that the series $\sum_{n=1}^{\infty}(-1)^{n-1}\frac{x^n}{n}$, $x \in [0,1]$ converges uniformly but not absolutely on $[0,1]$.

For each $n \in \mathbb{N}$, let

$$b_n(x) = \frac{x^n}{n}, \; x \in [0,1].$$

Then

$$0 \leq b_{n+1}(x) = \frac{x^{n+1}}{n+1} \leq \frac{x^{n+1}}{n} \leq \frac{x^n}{n} = b_n(x)$$

for all $x \in [0,1]$ and $n \in \mathbb{N}$. Furthermore,

$$\lim_{n\to\infty} b_n(x) = 0 \text{ for all } x \in [0,1].$$

Therefore, by the alternating series test, $\sum_{n=1}^{\infty}(-1)^{n-1}\frac{x^n}{n}$ converges for all $x \in [0,1]$. Let

$$s(x) = \sum_{n=1}^{\infty}(-1)^{n-1}\frac{x^n}{n}, \; s_n(x) = \sum_{k=1}^{n}(-1)^{k-1}\frac{x^k}{k}; \; x \in [0,1] \text{ and } n \in \mathbb{N}.$$

If $\varepsilon > 0$ is arbitrary and $\left\lceil - \right\rceil$ then for all $n \geq n_0$ and $x \in [0,1]$,

$$|s_n(x) - s(x)| \leq b_{n+1}(x) \leq \frac{1}{n+1} < \varepsilon.$$

Hence, $\sum_{n=1}^{\infty}(-1)^{n-1}\frac{x^n}{n}$ converges uniformly to s on $[0,1]$. However, for $x = 1$, the series becomes: $\sum_{n=1}^{\infty}\frac{(-1)^{n-1}}{n}$, and this series does not converge absolutely.

6.2A1 Assignment: Use the Weierstrass M-Test to show that the following series converges uniformly: (a) $\sum_{n=1}^{\infty}\frac{1}{n^2+x}$, $x \in [0,\infty)$ (b) $\sum_{n=1}^{\infty}\frac{x^2}{e^{nx}}$, $x \in [1,\infty)$.

6.2E4 Example: Show that $\sum_{n=1}^{\infty}\frac{x^n}{e^{nx}}$, $x\in[0,\infty)$ converges uniformly.

Let $f(x)=xe^{-x}$, $x\geq 0$. Then $f'(x)=e^{-x}-xe^{-x}=0\Rightarrow x=1$. If $x\in[0,1)$, then by the MVT, there exists $c\in(x,1)$ such that

$$f(1)-f(x)=(1-x)f'(c)=(1-c)e^{-c}>0.$$

Therefore, $f(1)>f(x)$, $0\leq x<1$. If $x\in(1,\infty)$, then there exists $c\in(1,x)$ such that

$$f(1)-f(x)=(1-x)f'(c).$$

Therefore, $f(1)>f(x)$, $x>1$ because $f'(c)=(1-c)e^{-c}<0$. Hence,

$$f(x)=xe^{-x}\leq f(1)=e^{-1}\leq 0.5 \text{ for all } x\in[0,\infty).$$

Thus

$$0\leq x^n e^{-nx}\leq(0.5)^n \text{ for all } n\in\mathbb{N} \text{ and } x\in[0,\infty).$$

Hence, by the Weierstrass M-Test, $\sum_{n=1}^{\infty}\frac{x^n}{e^{nx}}$ converges uniformly on $x\in[0,\infty)$, Since $\sum_{n=1}^{\infty}(0.5)^n$ converges.

6.2E5 Example: Let $s(x)=\begin{cases}0, & x=0\\ 1, & x\neq 0.\end{cases}$ Show that $\sum_{n=1}^{\infty}\frac{x^2}{(1+x^2)^n}$ converges (absolutely) to s on $\mathbb{R}$ but is not uniformly convergent on any interval containing the origin.

Let $x\in\mathbb{R}\setminus\{0\}$. Then the series is a geometric series with $|r|=(1+x^2)^{-1}<0$ and converges to

$$\frac{a}{1-r}=\frac{x^2/(1+x^2)}{1-1/(1+x^2)}=1.$$

If $x=0$, then each term in the series is 0 and the series converges to 0. Therefore, $\sum_{n=1}^{\infty}\frac{x^2}{(1+x^2)^n}$ converges (absolutely) to s on $\mathbb{R}$. Since each term in the series is continuous on $\mathbb{R}$, the partial sums, $s_n=\sum_{k=1}^{n}\frac{x^2}{(1+x^2)^k}$, $n\in\mathbb{N}$, are continuous on $\mathbb{R}$. However, s is not continuous on any interval containing the origin. If the convergence is uniform, then by (6.1T3C), $s=\lim_{n\to\infty}s_n$ should be continuous. It follows that the convergence is not uniform on any interval containing the origin.

6.2A2 Assignment:

(a) Show that $\sum_{n=1}^{\infty}\frac{(-1)^{n-1}}{n+x}$ converges uniformly but not absolutely on $[0,\infty)$.

(b) Suppose $\sum_{n=1}^{\infty}a_n$ converges absolutely. Prove that $\sum_{n=1}^{\infty}a_n x^n$ converges on $[-1,1]$.

(c) Suppose $\sum_{n=1}^{\infty}a_n$ converges. Prove that $\sum_{n=1}^{\infty}a_n x^n$ converges uniformly on $[0,1]$.

(d) Prove that $\sum_{n=2}^{\infty}\frac{x^n}{n(\log n)}$ converges uniformly on $[0,1]$.

6.2T3 Theorem:

For each $n\in\mathbb{N}$, let $f_n : E\to\mathbb{R}$ be continuous and suppose $\sum_{n=1}^{\infty}f_n$ converges uniformly to s on E, where $E\subseteq\mathbb{R}$. Then s is continuous on E and

$$\lim_{x\to p}\sum_{n=1}^{\infty}f_n(x)=s(p).$$

Proof: Apply (6.1T3C) to the sequence of partial sums $\{s_n\}: s(x)=\lim_{n\to\infty}s_n(x),\ x\in E$ is continuous and

$$\lim_{x\to p}\sum_{n=1}^{\infty}f_n(x)=\sum_{n=1}^{\infty}\lim_{x\to p}f_n(x)$$
$$=\sum_{n=1}^{\infty}f_n(p)=s(p).$$

6.2E6 Example: Show that the series $\sum_{n=1}^{\infty}x(1-x)^{n-1}$ does not converge uniformly on $[0,1]$.

If $x=0$, then each term of the series is 0, and it converges to 0. If $0<x\le 1$, then $|1-x|<1$ and the series converges to

$$\frac{x}{1-(1-x)}=1.$$

The terms of the series are continuous on $[0,1]$, and the series converges pointwise to the discontinuous function

$$s(x)=\begin{cases}0, & x=0\\ 1, & 0<x\le 1\end{cases}.$$

Therefore, the convergence cannot be uniform by (6.2T3).

6.2T4 Theorem:

Suppose for all $n \in \mathbb{N}$, $f_n \in \Re[a,b]$, and $\sum_{n=1}^{\infty} f_n$ converges uniformly to s on $[a,b]$, where $a,b \in \mathbb{R}$, $a < b$. Then $s \in \Re[a,b]$ and

$$\int_a^b s = \int_a^b \sum_{n=1}^{\infty} f_n = \sum_{n=1}^{\infty} \int_a^b f_n.$$

Proof: Apply (6.1T8) to the uniformly convergent sequence of partial sums $\left\{ s_n = \sum_{k=1}^{n} f_k \right\}$: $\sum_{n=1}^{\infty} f_n = s \in \Re[a,b]$ and

$$\int_a^b s = \int_a^b \lim_{n\to\infty} \sum_{k=1}^{n} f_k = \lim_{n\to\infty} \int_a^b \sum_{k=1}^{n} f_k = \lim_{n\to\infty} \sum_{k=1}^{n} \int_a^b f_k = \sum_{n=1}^{\infty} \int_a^b f_n.$$

6.2E7 Example: Prove that the series $\sum_{n=1}^{\infty} \frac{g(nx)}{n^2}$ is uniformly convergent on $[0,1]$, where $g(x) = x - [x]$, $x \in \mathbb{R}$ and find an expression for $\int_0^1 \left[\sum_{n=1}^{\infty} \frac{g(nx)}{n^2} \right] dx$.

For each $n \in \mathbb{N}$, define

$$f_n(x) = \frac{g(nx)}{n^2}, \; x \in [0,1].$$

Then

$$|f_n(x)| \le \frac{1}{n^2}, \text{ for all } x \in [0,1]$$

and $f_n \in \Re[a,b]$, $n \in [a,b]$. Therefore, by the Weierstrass M-Test, $\sum_{n=1}^{\infty} f_n$ converges uniformly on $[0,1]$, since $\sum_{n=1}^{\infty} \frac{1}{n^2}$ converges. Furthermore, for each $n \in \mathbb{N}$, $f_n \in \Re[0,1]$, because it is piecewise continuous. Therefore,

$$\int_0^1 [nx]dx = \sum_{k=1}^{n} \int_{(k-1)/n}^{k/n} [nx]dx$$

$$= \sum_{k=1}^{n} \int_{(k-1)/n}^{k/n} (k-1)dx$$

$$= \frac{1}{n} \cdot \frac{(n-1)n}{2} = \frac{n-1}{2}$$

and

$$\int_0^1 nxdx = \frac{n}{2}.$$

Thus

$$\int_0^1 f_n = \frac{1}{n^2}\left[\frac{n}{2}-\frac{n-1}{2}\right]$$
$$= \frac{1}{2n^2},\ n \in \mathbb{N}.$$

Hence,

$$\int_0^1\left[\sum_{n=1}^{\infty} f_n(x)\right]dx = \sum_{n=1}^{\infty}\int_0^1 f_n(x)\,dx$$
$$= \sum_{n=1}^{\infty}\frac{1}{2n^2}.$$

6.2A3 Assignment: Suppose $\sum_{n=1}^{\infty} a_n$ is absolutely convergent, where $a_n \in \mathbb{R}$ for all $n \in \mathbb{N}$. Prove that $\sum_{n=1}^{\infty} a_n x^{n-1}$ is uniformly convergent on $[0,1]$ and deduce that $\int_0^1 \sum_{n=1}^{\infty} a_n x^{n-1} = \sum_{n=1}^{\infty}\frac{a_n}{n}.$

6.2T5 Theorem:

Let $\{f_n\}$ be a sequence of differentiable functions on $[a,b]$ and suppose $\sum_{n=1}^{\infty} f_n(x_0)$ converges for some $x_0 \in [a,b]$. If $\sum_{n=1}^{\infty} f_n'$ converges uniformly on $[a,b]$, then $\sum_{n=1}^{\infty} f_n$ converges uniformly on $[a,b]$ and

$$\frac{d}{dx}\sum_{n=1}^{\infty} f_n(x) = \sum_{n=1}^{\infty} f_n'(x),\ x \in [a,b].$$

Proof: Apply (6.1T9) to the sequence of partial sums $\{s_n\}$: $\lim_{n\to\infty} s_n(x_0)$ exists and $\{s_n'\}$ converges uniformly on $[a,b]$. Therefore, $\sum_{n=1}^{\infty} f_n$ converges uniformly on $[a,b]$ and

$$\frac{d}{dx}\sum_{n=1}^{\infty} f_n(x) = \frac{d}{dx}\lim_{n\to\infty} s_n(x)$$
$$= \lim_{n\to\infty}\frac{d}{dx}s_n(x)$$
$$= \sum_{n=1}^{\infty} f_n'(x),\ x \in [a,b].$$

6.2E8 Example: Let $g:\mathbb{R}\to\mathbb{R}$ be defined by

$$g(x) = |x|,\ x \in [-1,1] \text{ and } g(x+2) = g(x) \text{ for all } x \in \mathbb{R}.$$

Then it is easily seen that

$$|g(x)-g(y)| \le |x-y| \text{ for all } x, y \in \mathbb{R}.$$

Furthermore, g is continuous. Define $f:\mathbb{R}\to\mathbb{R}$ by

$$f(x)=\sum_{n=0}^{\infty}\left(\frac{3}{4}\right)^n g\left(4^n x\right),\ x\in\mathbb{R}.$$

Then the above series converges uniformly on $\mathbb{R}$ because

$$\left(\frac{3}{4}\right)^n g\left(4^n x\right) \le \left(\frac{3}{4}\right)^n \text{ for all } x\in\mathbb{R} \text{ and } n\in\mathbb{N}$$

and $\sum_{n=0}^{\infty}\left(\frac{3}{4}\right)^n$ converges. Therefore, by (6.2T3), f is continuous. For $x\in\mathbb{R}$ and $m\in\mathbb{N}$, let $\delta_m = \pm\frac{1}{2}\cdot 4^{-m}$, where the sign is chosen so that no integer lies between $4^m x$ and $4^m\left(x+\delta_m\right)$. This is possible because $4^m\left|\delta_m\right| = \frac{1}{2}$. Now define

$$h_n = \frac{g\left(4^n\left(x+\delta_m\right)\right)-g\left(4^n x\right)}{\delta_m}.$$

If $n>m$, then $4^n\delta_m$ is an even integer, so that $h_n=0$. If $0\le n\le m$, then

$$|g(x)-g(y)|\le|x-y| \text{ for all } x,y\in\mathbb{R} \Rightarrow \left|h_n\right|\le 4^n.$$

Therefore

$$\left|\frac{f\left(x+\delta_m\right)-f(x)}{\delta_m}\right| = \left|\sum_{n=0}^{m}\left(\frac{3}{4}\right)^n h_n\right|$$

$$\ge \left(\frac{3}{4}\right)^m 4^m - \sum_{n=0}^{m-1}\left(\frac{3}{4}\right)^n 4^n$$

$$= 3^m - \sum_{n=0}^{m-1} 3^n$$

$$= 3^m - \frac{3^m-1}{3-1} = \frac{3^m+1}{2}$$

because $\left|h_m\right| = 4^m$. Since $\lim_{m\to\infty}\delta_m = 0$,

$$\lim_{m\to\infty}\frac{f\left(x+\delta_m\right)-f(x)}{\delta_m} = f'(x),$$

if f is differentiable at x. However, $\lim\limits_{m\to\infty}\frac{3^m+1}{2}=\infty$. Therefore, f is not differentiable at x. **This is an example of a continuous function that is nowhere differentiable.**

6.2A4 Assignment: Let $E=(0,\infty)$ and $f_n(x)=(1+nx)^{-2}$ for all $n\in\mathbb{N}$, $x\in E$. Use (6.2T5) to prove that $\sum\limits_{n=1}^{\infty}f_n'(x)=\frac{d}{dx}\sum\limits_{n=1}^{\infty}f_n(x)$ for all $x\in E$. (Hint: Weierstrass M-Test.)

6.2E9 Example: Show that the series $\sum\limits_{n=0}^{\infty}x^n$ converges on $(-1,1)$ and that the derivative of the sum can be obtained by term by term differentiation of the series.

Let $x\in(-1,1)$. Then

$$s_n(x)=\sum_{k=0}^{n}x^k=\frac{1-x^{n+1}}{1-x}\to\frac{1}{1-x}\text{ as } n\to\infty.$$

Therefore $\sum\limits_{n=0}^{\infty}x^n$ converges pointwise to $s(x)=\frac{1}{1-x}$ on $(-1,1)$. We have to show that $s'(x)=\sum\limits_{n=1}^{\infty}nx^{n-1}$ for all $x\in(-1,1)$ and $\sum\limits_{n=1}^{\infty}nx^{n-1}$ converges on $(-1,1)$. Let $x\in(-1,1)$. Then

$$\sum_{k=1}^{n}kx^{k-1}=\frac{1-x^n}{(1-x)^2}-\frac{nx^n}{1-x}\text{ and }\lim_{n\to\infty}\sum_{k=1}^{n}kx^{k-1}=\frac{1}{(1-x)^2}$$

because for $x\in(-1,1)$, $\lim\limits_{n\to\infty}nx^n=0$. Let $r=(|x|+1)/2<1$. Since $\lim\limits_{n\to\infty}nr^n=0$, there exists $n_0\in\mathbb{N}$ such that $|nr^n|<1$ for all $n\ge n_0$. Then

$$\left|nt^{n-1}\right|\le n|x|^{n-1}\le\frac{1}{r}\left(\frac{|x|}{r}\right)^{n-1}=\frac{1}{r}p^{n-1}\text{ for all } n\ge n_0 \text{ and } t\in\left[-|x|,|x|\right].$$

Let

$$M_n=n_0\text{ for } n=1,2,\dots,n_0-1\text{ and } M_n=\frac{1}{r}p^{n-1}\text{ for } n\ge n_0.$$

Then

$$\left|nt^{n-1}\right|\le M_n\text{ for all } n\in\mathbb{N}$$

and since $0<p<1$, $\sum\limits_{n=1}^{\infty}M_n$ converges. Hence, by the Weierstrass M-Test, $\sum\limits_{n=1}^{\infty}nt^{n-1}$ converges uniformly on $\left[-|x|,|x|\right]$. Therefore by (6.2T5),

$$s'(t)=\sum_{n=1}^{\infty}nt^{n-1}\text{ for all } t\in\left[-|x|,|x|\right].$$

In particular, $s'(x)=\sum_{n=1}^{\infty} nx^{n-1}$.

6.2T6 Theorem: Stone-Weierstrass Theorem

If $f:[a,b]\to\mathbb{R}$ is continuous, then there exists a sequence of polynomials $\{P_n\}$ such that $\{P_n\}$ converges uniformly to *f* on $[a,b]$.

Proof: Let $g:[0,1]\to\mathbb{R}$ be defined by

$$g(x)=f((1-x)a+xb),\ x\in[0,1]$$

and $h:[0,1]\to\mathbb{R}$ be defined by

$$h(x)=g(x)-g(0)-x[g(1)-g(0)],\ x\in[a,b].$$

Then

$$g(0)=f(a),\ g(1)=f(b),\ \text{and } h(0)=h(1)=0.$$

Since $g-h$ is a polynomial, if the result holds for *h*, then it holds for *g* as well. Furthermore, since

$$((1-x)a+xb)\,p_n=q_{n+1},$$

where p_n and q_{n+1} are polynomials, if the result holds for *g*, then it holds for *f* as well. Therefore, without loss of generality, assume that $f:[0,1]\to\mathbb{R}$ and $f(0)=f(1)=0$. Furthermore, if we define

$$f(x)=0,\ x\in\mathbb{R}\setminus[0,1],$$

then *f* is uniformly continuous on $\mathbb{R}$. Define

$$Q_n(x)=c_n\left(1-x^2\right)^n$$

where $c_n=\left[\int_{-1}^{1}\left(1-x^2\right)^n dx\right]^{-1}$, for all $n\in\mathbb{N}$.

Since,

$$\begin{aligned}\int_{-1}^{1}\left(1-x^2\right)^n dx &= 2\int_{0}^{1}\left(1-x^2\right)^n dx\\ &\geq 2\int_{0}^{1/\sqrt{n}}\left(1-nx^2\right)dx\\ &=\frac{4}{3\sqrt{n}}>\frac{1}{\sqrt{n}},\end{aligned}$$

we have

$$c_n<\sqrt{n}\ \text{for all } n\in\mathbb{N}.$$

The inequality $\left(1-x^2\right)^n\geq 1-nx^2$ follows by applying the MVT to the function *g* defined by

$$g(x)=\left(1-x^2\right)^n-\left(1-nx^2\right),\ x\in[0,1].$$

Let $\varepsilon > 0$ be given. Then there exists $\delta \in (0,1)$ such that

$$|f(x) - f(y)| < \frac{\varepsilon}{2} \text{ for all } |x - y| < \delta \text{ and } x, y \in \mathbb{R}$$

because f is uniformly continuous. Therefore,

$$\begin{aligned} 0 \le |Q_n(x)| &= Q_n(x) \\ &= c_n(1 - x^2)^n \\ &\le \sqrt{n}(1 - x^2)^n \\ &\le \sqrt{n}(1 - \delta^2)^n \end{aligned}$$

for all $x \in [-1,1]$ such that $\delta \le |x| \le 1$. The function f is clearly bounded; let $M = \sup_{\mathbb{R}} |f|$. Since $\lim_{n\to\infty} 4M\sqrt{n}(1-\delta^2)^n = 0$, there exists $n_0 \in \mathbb{N}$ such that

$$4M\sqrt{n}(1-\delta^2)^n < \frac{\varepsilon}{2} \text{ for all } n \ge n_0.$$

Define

$$P_n(x) = \int_{-1}^{1} f(x+t) Q_n(t)\,dt, \; 0 \le x \le 1.$$

Now $f(x+t) = 0,\ t \in \mathbb{R} \setminus [-x, 1-x]$, since $f(x) = 0,\ x \in \mathbb{R} \setminus [0,1]$. Thus,

$$\begin{aligned} P_n(x) &= \int_{-x}^{1-x} f(x+t) Q_n(t)\,dt \\ &= \int_{0}^{1} f(t) Q_n(t-x)\,dt, \; x \in (0,1), \end{aligned}$$

using a simple change of variables. The last integral clearly gives us a polynomial in x. Thus $\{P_n\}$ is a sequence of polynomials. Let $f_{(x)}(t) = f(x+t) - f(x);\ x, t, x+t \in [0,1]$. Since

$$\int_{-1}^{1} Q_n(x)\,dx = \int_{-1}^{1} c_n(1-x^2)^n\,dx = 1,$$

using the above inequalities, we obtain

$$\begin{aligned} |P_n(x) - f(x)| &= \left| \int_{-1}^{1} [f(x+t) - f(x)] Q_n(t)\,dt \right| \\ &\le \int_{-1}^{1} |f_{(x)}(t)| Q_n(t)\,dt \\ &= \int_{-1}^{-\delta} |f_{(x)}(t)| Q_n(t)\,dt + \int_{-\delta}^{\delta} |f_{(x)}(t)| Q_n(t)\,dt + \int_{\delta}^{1} |f_{(x)}(t)| Q_n(t)\,dt \\ &\le \int_{-1}^{-\delta} 2MQ_n(t)\,dt + \int_{-1}^{1} \frac{\varepsilon}{2} Q_n(t)\,dt + \int_{\delta}^{1} 2MQ_n(t)\,dt \\ &\le 4M\sqrt{n}(1-\delta^2)^n + \frac{\varepsilon}{2} \\ &< \frac{\varepsilon}{2} + \frac{\varepsilon}{2} = \varepsilon \end{aligned}$$

for all $n \geq n_0$ and $x \in [0,1]$. Hence, $\{P_n\}$ converges to f uniformly on $[0,1]$.

6.2T6C Corollary:

Suppose $f:[a,b] \to \mathbb{R}$ is continuous. If $\varepsilon > 0$ is given, then there exists a polynomial P such that

$$|f(x) - P(x)| < \varepsilon \text{ for all } x \in [a,b].$$

Proof: Let $\{P_n\}$ be the sequence of polynomials in (6.2T6). Take $P = P_{n_0}$.

6.2E10 Example: Let $f:[0,1] \to \mathbb{R}$ be continuous and

$$\int_0^1 x^{n-1} f(x)\,dx = 0 \text{ for all } n \in \mathbb{N}.$$

Prove that $f(x) = 0$ for all $x \in [0,1]$.

Since f is continuous, there exists $M > 0$ such that

$$|f(x)| \leq M \text{ for all } x \in [0,1].$$

Let $\varepsilon > 0$ be given. Then there exists a polynomial P such that

$$|f(x) - P(x)| < \frac{\varepsilon}{2M} = \varepsilon' \text{ for all } x \in [0,1].$$

Therefore,

$$\left|\int_0^1 \left[f(x)P(x) - \{f(x)\}^2\right] dx\right| \leq \int_0^1 |f(x)||P(x) - f(x)|\,dx$$

$$\leq \int_0^1 M\varepsilon'\,dx < \varepsilon.$$

If $P(x) = \sum_{k=0}^{n} a_k x^k$, then

$$\int_0^1 f(x)P(x)\,dx = \int_0^1 f(x)\sum_{k=0}^{n} a_k x^k\,dx = \sum_{k=0}^{n} a_k \int_0^1 x^k f(x)\,dx = 0.$$

It follows that

$$0 \leq \int_0^1 f^2(x)\,dx$$

$$= \left|\int_0^1 \left[0 - f^2(x)\right] dx\right|$$

$$= \left|\int_0^1 \left[f(x)P(x) - \{f(x)\}^2\right] dx\right| < \varepsilon.$$

Since $\varepsilon > 0$ was arbitrary,

$$\int_0^1 f^2(x)\,dx = 0.$$

Therefore,

$$f^2(x)=0 \text{ for all } x\in[0,1]$$

because f^2 is nonnegative and continuous on $[0,1]$. Hence, $f(x)=0$ for all $x\in[0,1]$.

6.2E11 Example: Let $f(x)=|x|$, $x\in[-1,1]$. Show that if $\varepsilon>0$ is given, then there exists a polynomial P such that

$$P(0)=0 \text{ and } |f(x)-P(x)|<\varepsilon \text{ for all } x\in[-1,1].$$

If $\varepsilon>0$ is given, then there exists a polynomial $Q(x)$ such that

$$|f(x)-Q(x)|<\frac{\varepsilon}{2} \text{ for all } x\in[-1,1].$$

Let

$$P(x)=Q(x)-Q(0).$$

Then $P(0)=0$ and for all $x\in[-1,1]$,

$$\begin{aligned}|f(x)-P(x)| &\le |f(x)-Q(x)|+|Q(x)-P(x)|\\ &=|f(x)-Q(x)|+|Q(0)-f(0)|\\ &<\frac{\varepsilon}{2}+\frac{\varepsilon}{2}=\varepsilon.\end{aligned}$$

6.2A5 Assignment: Suppose $f:[0,1]\to\mathbb{R}$ is continuous. Prove that if $\varepsilon>0$ is arbitrary, then there exists a polynomial P with rational coefficients such that

$$|f(x)-P(x)|<\varepsilon \text{ for all } x\in[a,b].$$

Hint: Use (6.2T6C) to find a polynomial Q of degree n with real coefficients to approximate f to within $\frac{\varepsilon}{2}$. Then use the fact that rational numbers are dense in $\mathbb{R}$ to obtain a polynomial P with rational coefficients so that Q approximates P to within $\frac{\varepsilon}{2}$.

6.3 Power Series and Special Functions

In this section, we shall study real power series, formally define trigonometric and hyperbolic functions, and obtain some of their basic properties.

6.3D1 Definition:

Let $\{c_n\}_{n=0}^{\infty}$ be a real sequence and $a\in\mathbb{R}$. A series of the form

$$\sum_{n=0}^{\infty} c_n(x-a)^n,\ x\in E\subseteq\mathbb{R}$$

is called a **power series centered** at a. The members of $\{c_n\}_{n=0}^{\infty}$ are called the **coefficients** of the power series. The set E will be determined by the sequence of coefficients $\{c_n\}_{n=0}^{\infty}$.

6.3T1 Theorem:

Let $\sum_{n=0}^{\infty} c_n (x-a)^n$ be a power series and let $L = \overline{\lim_{n\to\infty}} \sqrt[n]{|c_n|}$.

(a) If $L > 0$, then the series converges absolutely when $|x-a| < R = \frac{1}{L}$ and diverges when $|x-a| > R$.

(b) If $L = 0$, then the series converges absolutely on $\mathbb{R}$.

(c) If $L = \infty$, then the series converges only when $x = a$.

Proof: Let $a_n = c_n (x-a)^n$, $n \geq 0$. Then

$$\sqrt[n]{|a_n|} = \sqrt[n]{\left|c_n (x-a)^n\right|}$$
$$= \sqrt[n]{|c_n|}\,|x-a|.$$

(a) $\lim_{n\to\infty} \sqrt[n]{|a_n|} = L|x-a| < 1 \Leftrightarrow |x-a| < \frac{1}{L} = R$. The result follows using (2.8T5a).

(b) $\lim_{n\to\infty} \sqrt[n]{|a_n|} = L|x-a| = 0 < 1$. The result follows using (2.8T5b).

(c) $\lim_{n\to\infty} \sqrt[n]{|a_n|} = L|x-a| = \infty > 1$ if $x \neq a$ and $\lim_{n\to\infty} \sqrt[n]{|a_n|} = L|x-a| = 0$ if $x = a$. Hence result.

Note: If $\lim_{n\to\infty} \left|\frac{a_{n+1}}{a_n}\right|$ exists in $\mathbb{R} \cup \{\infty\}$, then $\lim_{n\to\infty} \left|\frac{a_{n+1}}{a_n}\right| = L$.

6.3D2 Definition: Radius of Convergence

Let $R = \frac{1}{L}$ as defined above with $R = 0$ if $L = \infty$ and $R = \infty$ if $L = 0$. Then R is called the **radius of convergence** of the series. If $R > 0$, the series converges absolutely on $(a-R, a+R)$. If $R = \infty$, then this interval is $\mathbb{R}$. The set of all points at which the series converges is called the **interval of convergence,** denoted by I. If $R = 0$, I is just a singleton. We will assume that I is **not trivial**, i.e., $\neq 0$. We may also assume that the center ***a* is 0.** In this case, the series converges absolutely on $(-R, R)$ or ∞. We define $f : I \to \mathbb{R}$ by

$$f(x) = \sum_{n=0}^{\infty} c_n (x-a)^n,\ x \in I.$$

6.3E1 Example: Find the radius and interval convergence of each power series:

(a) $\sum_{n=0}^{\infty} x^n$ (b) $\sum_{n=1}^{\infty} \frac{x^n}{n}$ (c) $\sum_{n=1}^{\infty} \frac{x^n}{n^2}$.

(a) $$s_n(x) = \frac{1-x^{n+1}}{1-x},\ x \neq 1 \text{ and } s_n(1) = (n+1).$$

Therefore, $\{s_n\}$ converges to $\dfrac{1}{1-x}$ if and only if $|x| < 1$. Hence, $R = 1$ and $I = (-1,1)$.

(b) $$\lim_{n\to\infty} \sqrt[n]{|c_n|} = \lim_{n\to\infty} \sqrt[n]{1/n} = 1.$$

Therefore, $R = 1$ and I contains $(-1,1)$. The series converges absolutely on $(-1,1)$. If $x = 1$, we have the harmonic series $\sum_{n=1}^{\infty} \frac{1}{n}$ that diverges. If $x = -1$, we have the alternating (harmonic) series that converges conditionally. Hence, $I = [-1,1]$. and the convergence is absolute only on $(-1,1)$.

(c) $$\lim_{n\to\infty} \sqrt[n]{|c_n|} = \lim_{n\to\infty} \sqrt[n]{1/n^2} = 1.$$

Therefore, $R = 1$ and I contains $(-1,1)$. The series converges absolutely on $(-1,1)$. If $x = \pm 1$, then $\sum_{n=1}^{\infty} \left|\frac{x^n}{n^2}\right| = \sum_{n=1}^{\infty} \frac{1}{n^2}$; the latter series converges. Hence $\sum_{n=1}^{\infty} \frac{x^n}{n^2}$ is absolutely convergent on its interval of convergence $I = [-1,1]$.

6.3A1 Find the radius of convergence and interval of convergence of each of the following power series:

(a) $$\sum_{n=1}^{\infty} \frac{n^{2n}}{(2n)!} x^n$$

(b) $$\sum_{n=0}^{\infty} \frac{(-1)^n}{2n+1} x^{2n+1}$$

6.3T1 Theorem:

Let $f(x) = \sum_{n=0}^{\infty} c_n x^n$ be a power series with radius of convergence $R > 0$. Then f is continuous on $(-R, R)$. Note: R may be ∞.

Proof: Let $x \in (-R, R)$. Then

$$|c_n t^n| = |c_n||t|^n \leq |c_n||x|^n$$

for all $n \in \mathbb{N}$ and all $t \in \mathbb{R}$ with $|t| \le |x|$. Since $\sum_{n=0}^{\infty} c_n x^n$ converges absolutely, $\sum_{n=0}^{\infty} c_n t^n$ converges absolutely and uniformly on $[-|x|,|x|]$, by the Weierstrass M-Test. Therefore, by (6.1T3C), f is continuous on $[-|x|,|x|]$. In particular f is continuous at x. Since x was arbitrary, f is continuous on $(-R, R)$.

6.3T2 Theorem: Abel's Theorem

Suppose the radius of convergence of $f(x) = \sum_{n=0}^{\infty} c_n x^n$ is $R = 1$. If $\sum_{n=0}^{\infty} c_n$ converges, then $\lim_{x \to 1-} f(x) = \sum_{n=0}^{\infty} c_n$.

Proof: Let

$$s_{-1} = 0,\ s_n = \sum_{k=0}^{n} a_k,\ n = 0,1,2,...$$

Then

$$\sum_{k=0}^{n} a_k x^k = \sum_{k=0}^{n-1} s_k \left(x^k - x^{k+1}\right) + s_n x^n - s_{-1} x^0$$

by (2.8L1). Therefore,

$$\sum_{k=0}^{n} a_k x^k = \sum_{k=0}^{n-1} s_k \left(x^k - x^{k+1}\right) + s_n x^n$$

$$= (1-x) \sum_{k=0}^{n-1} s_k x^k + s_n x^n.$$

By taking limits as $n \to \infty$, we obtain

$$f(x) = \lim_{n \to \infty} \sum_{k=0}^{n} a_k x^k$$

$$= (1-x) \lim_{n \to \infty} \sum_{k=0}^{n-1} s_k x^k + \lim_{n \to \infty} s_n x^n$$

$$= (1-x) \sum_{k=0}^{\infty} s_k x^k.$$

Note that for all $x \in (-1,1)$, $(1-x) \sum_{k=0}^{\infty} x^k = 1$. Let

$$\lim_{n \to \infty} s_n = \sum_{n=0}^{\infty} c_n = s.$$

If $\varepsilon > 0$ is given, then there exists $n_0 \in \mathbb{N}$ such that

$$|s_n - s| < \varepsilon / 2 \text{ for all } n \ge n_0.$$

Let $M = \sum_{k=0}^{n_0} |s_k - s|$. Then there exists $\delta \in (0,1)$ such that

$$2(1-x)M < \varepsilon \text{ whenever } 1-\delta < x < 1.$$

Therefore for all $x \in (1-\delta,1)\setminus\{1\}$, we have

$$\begin{aligned}
|f(x)-s| &= \left|(1-x)\sum_{k=0}^{\infty} s_k x^k - s\right| \\
&= \left|(1-x)\sum_{k=0}^{\infty} s_k x^k - s(1-x)\sum_{k=0}^{\infty} x^k\right| \\
&\le (1-x)\sum_{k=0}^{\infty} |s_k - s| x^k \\
&\le (1-x)\sum_{k=0}^{n_0} |s_k - s| + \frac{\varepsilon}{2}(1-x)\sum_{k=n_0+1}^{\infty} x^k \\
&\le (1-x)M + \frac{\varepsilon}{2}(1-x)\sum_{k=0}^{\infty} x^k \\
&< \frac{\varepsilon}{2} + \frac{\varepsilon}{2} = \varepsilon.
\end{aligned}$$

Hence, $\lim_{x\to 1-} f(x) = \sum_{n=0}^{\infty} c_n$.

6.3E2 Example: Show that $\log 2 = \sum_{n=1}^{\infty} \frac{(-1)^{n-1}}{n} = 1 - \frac{1}{2} + \frac{1}{3} - \frac{1}{4} + \ldots$

Consider the series

$$\sum_{n=0}^{\infty} (-1)^n t^n.$$

It is easily seen that the radius of convergence of this series is 1. Furthermore, since

$$s_n = \sum_{k=0}^{n} (-t)^k = \frac{1-(-t)^{n+1}}{1+t}, \; n \in \mathbb{N}, \; t \neq 1,$$

the series converges to

$$g(t) = \frac{1}{1+t}, \; t \in (-1,1).$$

If $x \in (-1,1)$, then the convergence is uniform on $[-|x|,|x|]$. Therefore, by (6.2T4),

$$\begin{aligned}
f(x) = \log(1+x) &= \int_0^x \frac{1}{1+t}\,dt = \sum_{n=0}^{\infty} (-1)^n \int_0^x t^n\,dt \\
&= \sum_{n=0}^{\infty} (-1)^n \frac{x^{n+1}}{n+1} = \sum_{n=1}^{\infty} \frac{(-1)^{n-1}}{n} x^n, \; x \in (-1,1).
\end{aligned}$$

The above series has radius of convergence 1 and also converges for $x=1$, i.e., the series of coefficients $\sum_{n=1}^{\infty}\frac{(-1)^{n-1}}{n}$ converges. Hence, by (6.3T2),

$$\log 2=\lim_{x\to 1-}f(x)=\sum_{n=1}^{\infty}\frac{(-1)^{n-1}}{n}.$$

6.3T3 Theorem:

Suppose the radius of convergence of $f(x)=\sum_{n=0}^{\infty}c_n x^n$ is *R*. Then

(a) $\int_0^x f=\sum_{n=0}^{\infty}\frac{c_n}{n+1}x^{n+1}=\sum_{n=1}^{\infty}\frac{c_{n-1}}{n}x^n,\ x\in(-R,R).$

(b) *f* is differentiable on $(-R,R)$ and

$$f'(x)=\sum_{n=1}^{\infty}nc_n x^{n-1}=\sum_{n=0}^{\infty}(n+1)c_{n+1}x^n,\ x\in(-R,R).$$

Proof: Let $x\in(-R,R)$. Then the series is uniformly convergent on $[-|x|,|x|]$.

(a) By (6.2T4),

$$\int_0^x f=\sum_{n=0}^{\infty}\int_0^x c_n t^n dt$$
$$=\sum_{n=0}^{\infty}\frac{c_n}{n+1}x^{n+1}.$$

(b) $\overline{\lim_{n\to\infty}}\sqrt[n]{|nc_n|}=\left[\overline{\lim_{n\to\infty}}\sqrt[n]{n}\right]\left[\overline{\lim_{n\to\infty}}\sqrt[n]{|c_n|}\right]=1\cdot R=R.$

Therefore, the series $\sum_{n=1}^{\infty}nc_n x^n$ has radius of convergence *R*. It follows that the radius of convergence of $\sum_{n=1}^{\infty}nc_n x^{n-1}$ is also *R*. Furthermore, the series is uniformly convergent on $[-|x|,|x|]$ if $x\in(-R,R)$. Therefore,

$$f'(t)=\sum_{n=1}^{\infty}nc_n t^{n-1},$$

for all $t\in[-|x|,|x|]$, by (6.2T5). In particular,

$$f'(x)=\sum_{n=1}^{\infty}nc_n x^{n-1}.$$

Hence, *f* is differentiable on $(-R,R)$ and

$$f'(x) = \sum_{n=1}^{\infty} nc_n x^{n-1}$$
$$= \sum_{n=0}^{\infty} (n+1)c_{n+1}x^n,\ x \in (-R, R).$$

6.3T3C1 Corollary:

Suppose the radius of convergence of $f(x) = \sum_{n=0}^{\infty} c_n x^n$ is *R*. Then *f* has derivatives of all orders in $(-R, R)$ and

$$f^{(k)}(x) = \sum_{n=k}^{\infty} n(n-1)...(n-k+1)c_n x^{n-k},\ k = 0,1,2,...$$

In particular,

$$f^{(k)}(0) = k!c_k,\ k = 0,1,2,...$$

Proof: $f^{(k)}(x) = \sum_{n=k}^{\infty} n(n-1)...(n-k+1)c_n x^{n-k},\ k = 0,1,2,...$

follows immediately, if we apply (6.3T3b) to each of the functions $f, f', f'',...$ If $x = 0$, then the series reduces to the first term:

$$f^{(k)}(0) = k(k-1)...(k-k+1)c_k = k!c_k,\ k = 0,1,2,...$$

Note: The function $f:(-R, R) \to \mathbb{R}$ is infinitely differentiable and belongs to $C^{\infty}(-R, R)$.

6.3T3C2 Corollary:

Suppose the radius of convergence of each function $f(x) = \sum_{n=0}^{\infty} c_n x^n$ and $g(x) = \sum_{n=0}^{\infty} d_n x^n$ is *R*. Then $f = g$ if and only if $c_n = d_n$ for $n = 0,1,2,...$

Proof: Clearly if $c_n = d_n$ for $n = 0,1,2,...,$ then $f = g$.

Conversely, if $f = g$, then $f^{(n)} = g^{(n)}$ for $n = 0,1,2,...$ In particular, $f^{(n)}(0) = g^{(n)}(0)$ and

$$c_n = \frac{f^{(n)}(0)}{n!} = \frac{g^{(n)}(0)}{n!} = d_n \text{ for } n = 0,1,2,...$$

by the previous corollary.

Note: We have seen that if a function is represented by a power series on $(-R, R)$, then it is in $C^{\infty}(-R, R)$. However, if $f \in C^{\infty}(-R, R)$, then it does not follow that *f* has a power series representation as can be seen by the following example.

6.3A2 Assignment: Use the power series expansion of $f(x) = \dfrac{1}{1-x}$ to evaluate the following:

(a) $\sum_{n=3}^{\infty} n^3 x$ (b) $\sum_{n=1}^{\infty} \frac{n}{2^n}.$

6.3E3 Example: Let $f:\mathbb{R}\to\mathbb{R}$ be defined by $f(x)=\begin{cases} e^{-1/x^2}, & x\neq 0 \\ 0, & x=0 \end{cases}$

Since $\lim_{x\to 0} e^{-1/x^2}=0$, f is continuous at 0.

$$f'(0)=\lim_{h\to 0}\frac{e^{-1/h^2}}{h}=\lim_{x\to\infty}\frac{x}{e^{x^2}}$$
$$=\lim_{x\to\infty}\frac{1}{2xe^{x^2}}=0$$

and

$$f'(x)=\frac{2}{x^3}e^{-1/x^2} \text{ if } x\neq 0.$$

Thus,

$$f'(x)=\begin{cases} \frac{2}{x^3}e^{-1/x^2}, & x\neq 0 \\ 0, & x=0. \end{cases}$$

Suppose for $n=0,1,2,\ldots,$

$$f^{(n)}(x)=\begin{cases} P_n(1/x^3)e^{-1/x^2}, & x\neq 0 \\ 0, & x=0, \end{cases}$$

where P_n is a polynomial of degree n. Then

$$\lim_{h\to 0}\frac{P_n(1/h^3)e^{-1/h^2}}{h}=\lim_{x\to\infty}\frac{x^3P_n(x^3)}{e^{x^2}}=\lim_{x\to\infty}\frac{Q_{n+1}(x^3)}{e^{x^2}}=0$$

by using L'Hospital's rule $3(n+1)$ times; Q_{n+1} is a polynomial of degree $n+1$. Therefore,

$$f^{(n+1)}(0)=0.$$

Suppose $x\neq 0$. Then

$$f^{(n+1)}(x)=-\frac{3}{x^4}P_n'(1/x^3)e^{-1/x^2}+P_n(1/x^3)\frac{2}{x^3}e^{-1/x^2}$$
$$=Q_{n+1}(1/x^3)e^{-1/x^2},$$

where Q_{n+1} is a polynomial of degree $n+1$. Hence by the modified principle of mathematical induction, for all $n=0,1,2,\ldots,$

$$f^{(n)}(x)=\begin{cases} P_n(1/x^3)e^{-1/x^2}, & x\neq 0 \\ 0, & x=0, \end{cases}$$

where P_n is a polynomial of degree *n*. if

$$f(x)=\sum_{n=0}^{\infty}c_n x^n,$$

then

$$c_n=\frac{f^{(n)}(0)}{n!}=0 \text{ for } n=0,1,2,\ldots$$

Since $f(x)\neq 0$ for all $x\neq 0$, this is impossible. Hence, *f* does not have a power series representation.

6.3D3 Definition: Taylor Polynomial

Let $f:I\to\mathbb{R}$, where *I* is an open interval in $\mathbb{R}$, $a\in I$, and $n\in\mathbb{N}$. Suppose $f^{(n)}$ exists on I. Then the polynomial of degree *n* defined by

$$T_n(f,a)(x)=\sum_{k=0}^{n}\frac{f^{(k)}(a)}{k!}(x-a)^k,\ x\in I$$

is called the **Taylor polynomial** of order *n* about the point *a*. If $f\in C^{\infty}(I)$, then the series

$$\sum_{n=0}^{\infty}\frac{f^{(n)}(a)}{n!}(x-a)^n$$

is called the **Taylor series** of *f* about *a*. For the special case $a=0$, the series is called a **Maclaurin series**. For each $n\in\mathbb{N}$, the function

$$\begin{aligned}R_n(x)&=R_n(f,a)(x)\\&=f(x)-T_n(f,a)(x),\ x\in I\end{aligned}$$

is called the **remainder** of f and T_n.

Note: For the function in (6.3E3), each coefficient of its Maclaurin series is 0. Therefore, the Maclaurin series reduces to 0 and is not the same as the function.

6.3T4 Theorem:

Suppose $f\in C^{\infty}(I)$, where *I* is an open interval in $\mathbb{R}$, $a\in I$. Then for all $x\in I$,

$$f(x)=\lim_{n\to\infty}T_n(f,a)(x)=\sum_{n=0}^{\infty}\frac{f^{(n)}(a)}{n!}(x-a)^n$$

if and only if

$$\lim_{n\to\infty}R_n(f,a)(x)=0.$$

Proof: Since

$$\begin{aligned}\lim_{n\to\infty}R_n(f,a)(x)&=\lim_{n\to\infty}\left[f(x)-T_n(f,a)(x)\right]\\&=f(x)-\lim_{n\to\infty}T_n(f,a)(x)\end{aligned}$$

for all $x \in I,$ the result follows.

6.3T5 Theorem: Extended Mean Value Theorem

Let $f: I \to \mathbb{R},$ where I is an open interval in $\mathbb{R},$ $a \in I,$ and $n \in \mathbb{N}.$ Suppose $f^{(n+1)}$ exists on I. Then for any $x \in I,$

$$f(x)=\sum_{k=0}^{n}\frac{f^{(k)}(a)}{k!}(x-a)^{k}+\frac{f^{(n+1)}(c)}{(n+1)!}(x-a)^{n+1},$$

where c is a number between x and a. The remainder

$$R_n(x)=R_n(f,a)(x)=\frac{f^{(n+1)}(c)}{(n+1)!}(x-a)^{n+1}$$

is called the **Lagrange form** of the remainder.

Proof: Let $x \in I.$ Without loss of generality, we may assume that $x > a.$ Let M be defined by

$$f(x)=\sum_{k=0}^{n}\frac{f^{(k)}(a)}{k!}(x-a)^{k}+M(x-a)^{n+1}.$$

The proof would be complete if we show that for some $c \in I,$

$$(n+1)!M = f^{(n+1)}(c).$$

Define g on I by

$$g(t)=f(t)-T_n(f,a)(t)-M(t-a)^{n+1},\ t \in I.$$

Since g is a polynomial of degree at most n,

$$g^{(n+1)}(t)=f^{(n+1)}(t)-(n+1)!M.$$

Furthermore,

$$g(a)=g'(a)=\ldots=g^{(n)}(a)=0,$$

since

$$T_n^{(k)}(f,a)(a)=f^{(k)}(a),\ k=0,1,\ldots,n.$$

By the choice of M,

$$g(x)=f(x)-T_n(f,a)(x)-M(x-a)^{n+1}=0.$$

First apply the MVT to g on $[a,x]$ to obtain

$$0=g(x)-g(a)=g'(x_1)(x-a) \text{ for some } x_1 \in (a,x).$$

Thus $g'(x_1)=0.$ Now apply the MVT to g' on $[a,x_1]$ to obtain

$$0=g'(x_1)-g'(a)=g''(x_2)(x_1-a) \text{ for some } x_2 \in (a,x_1).$$

Therefore, $g''(x_2)=0.$ Proceeding in this manner, we obtain

$$x_n \in (a, x_{n-1}) \text{ such that } g^{(n)}(x_n) = 0.$$

Apply the MVT to $g^{(n)}$ on $[a, x_n]$ to obtain $c \in (a, x_n)$ such that

$$0 = g^{(n)}(x_n) - g^{(n)}(a) = g^{(n+1)}(c)(x_n - a).$$

Thus

$$g^{(n+1)}(c)(x_n - a) = 0, \text{ i.e., } 0 = g^{(n+1)}(c) = f^{(n+1)}(c) - (n+1)!M$$

for some c between x and a. If $x = a$, then the series reduces to $f(a)$ and the result holds.

6.3E4 Example: Find the Maclaurin series for the function $f(x) = e^x$, $x \in \mathbb{R}$ and show that the series converges to f on $\mathbb{R}$.

$$f(x) = e^x \Rightarrow f^{(n)}(x) \text{ for all } n = 0, 1, 2, \ldots \text{ and } x \in \mathbb{R}.$$

Therefore,

$$f^{(n)}(0) = 1 \text{ for all}$$

and the Maclaurin series of f is

$$\sum_{n=0}^{\infty} \frac{1}{n!} x^n, \; x \in \mathbb{R}.$$

If $x = 0$, then the series reduces to

$$1 = e^0 - f(0).$$

Suppose $x < 0$. Then

$$\left| R_n(f,0)(x) \right| = \left| \frac{f^{(n+1)}(c)}{(n+1)!} x^{n+1} \right|$$

$$= \left| \frac{e^c}{(n+1)!} x^{n+1} \right| \leq \frac{\left| x^{n+1} \right|}{(n+1)!} \to 0$$

as $n \to \infty$ because $\sum_{n=0}^{\infty} \frac{x^{n+1}}{(n+1)!}$, $x \in \mathbb{R}$ is absolutely convergent. Suppose $x > 0$. Then

$$0 \leq R_n(f,0)(x) = \frac{e^c x^{n+1}}{(n+1)!} \leq e^x \frac{x^{n+1}}{(n+1)!} \to 0$$

as $n \to \infty$ because $\sum_{n=0}^{\infty} \frac{x^{n+1}}{(n+1)!}$, $x \in \mathbb{R}$ is absolutely convergent and e^x is fixed. Hence,

$$e^x = \sum_{n=0}^{\infty} \frac{1}{n!} x^n, \; x \in \mathbb{R}.$$

6.3T6 Theorem:

If $f:\mathbb{R}\to\mathbb{R}$ is twice differentiable and satisfies $f''(x)=-f(x)$ for all $x\in\mathbb{R}$, then

$$f(x)=c_1+c_2x-\frac{c_1}{2!}x^2-\frac{c_2}{3!}x^3+\frac{c_1}{4!}x^4+\frac{c_2}{5!}x^5-\ldots \text{ for all } x\in\mathbb{R}.$$

Proof: Since $f''=-f,\ f''$ must also be twice differentiable. Therefore, it follows that *f* has derivatives of all orders on $\mathbb{R}$. Furthermore,

$$f'''=-f'\Rightarrow f^{(5)}=-f'''\Rightarrow \ldots f^{(2n+1)}=(-1)^n f' \text{ for all } n\in\mathbb{N}.$$

We also have

$$f^{(4)}=-f''\Rightarrow f^{(6)}=-f^{(4)}\Rightarrow \ldots f^{(2n)}=(-1)^n f \text{ for all } n\in\mathbb{N}.$$

Therefore, the Maclaurin series of *f* is

$$f(0)+\frac{f'(0)}{1!}x-\frac{f(0)}{2!}x^2-\frac{f'(0)}{3!}x^3+\frac{f(0)}{4!}x^4+\frac{f'(0)}{5!}x^5-\ldots,\ x\in\mathbb{R}.$$

Let

$$c_1=f(0) \text{ and } c_2=f'(0).$$

If $x=0$, the series reduces to $f(0)$. Suppose $x\in\mathbb{R}\setminus\{0\}$. Then

$$\left|R_n(f,0)(x)\right|=\left|\frac{f^{(n+1)}(c)}{(n+1)!}x^{n+1}\right|$$

$$=\frac{\left|f^{(n+1)}(c)\right|\left|x^{n+1}\right|}{(n+1)!}\le M\frac{\left|x^{n+1}\right|}{(n+1)!}$$

where c is between x and 0 and $M=\max\left\{\max_{[-|x|,|x|]}|f|,\max_{[-|x|,|x|]}|f'|\right\}$. Note that f and all of its derivatives are continuous and therefore bounded on compact sets. Because $\sum_{n=0}^{\infty}\frac{x^{n+1}}{(n+1)!},\ x\in\mathbb{R}$ is absolutely convergent,

$$M\frac{\left|x^{n+1}\right|}{(n+1)!}\to 0 \text{ as } n\to\infty.$$

The result follows now by (6.3T5).

6.3D4 Definition: Trigonometric Functions Sine and Cosine

(a) The **sine** function, denoted by sin, is defined for all $x\in\mathbb{R}$ and is given by

$$\sin x=x-\frac{x^3}{3!}+\frac{x^5}{5!}-\ldots=\sum_{n=0}^{\infty}\frac{(-1)^n}{(2n+1)!}x^{2n+1},\ x\in\mathbb{R}.$$

(b) The **cosine** function, denoted by cos, is defined for all $x\in\mathbb{R}$ and is given by

$$\cos x=1-\frac{x^2}{2!}+\frac{x^4}{4!}-\ldots=\sum_{n=0}^{\infty}\frac{(-1)^n}{(2n)!}x^{2n},\ x\in\mathbb{R}.$$

Note: The function f in (6.3T6) can be expressed as $f(x)=c_1\cos x+c_2\sin x,\ x\in\mathbb{R}$. It is quite clear that the power series for sin and cos converge absolutely on $\mathbb{R}$. It follows that their derivatives are given by

$$\frac{d}{dx}\sin x=\cos x,\quad \frac{d}{dx}\cos x=-\sin x,\ x\in\mathbb{R}.$$

Therefore, both sin and cos satisfy

$$f''=-f \text{ on } \mathbb{R}.$$

Furthermore, $\sin 0=0$ and $\cos 0=1$. From the series representations, we obtain,

$$\sin(-x)=-\sin x \text{ and } \cos(-x)=\cos x.$$

Notation: $(\sin x)^n$ is denoted by $\sin^n x$ and $(\cos x)^n$ is denoted by $\cos^n x$ for $x\in\mathbb{R},\ n\in\mathbb{N}$.

6.3T7 Theorem:

For all $x,y\in\mathbb{R}$ the following identities hold:

(a) $\sin^2 x+\cos^2 x=1,$

(b) $\sin(x+y)=\sin x\cos y+\cos x\sin y,$

(c) $\cos(x+y)=\cos x\cos y-\sin x\sin y.$

Proof:

(a) $\dfrac{d}{dx}\left(\sin^2 x+\cos^2 x\right)=2\sin x\cos x+2\cos x(-\sin x)=0,\ x\in\mathbb{R}.$

Therefore, $\sin^2 x+\cos^2 x=\text{constant}$ on $\mathbb{R}$. But $\sin^2 0+\cos^2 0=0+1=1$. Hence result.

(b), (c) Fix $y\in\mathbb{R}$. Then

$$\frac{d}{dx}\sin(x+y)=\cos(x+y) \text{ and } \frac{d}{dx}\cos(x+y)=-\sin(x+y).$$

Thus

$$\frac{d^2}{dx^2}\sin(x+y)=-\sin(x+y)$$

and $f(x)=\sin(x+y),\ x\in\mathbb{R}$ satisfies $f''=-f$ on $\mathbb{R}$. Therefore,

$$\sin(x+y)=c_1\cos x+c_2\sin x,\ x\in\mathbb{R}$$

for some $c_1,c_2\in\mathbb{R}$. Differentiating the above with respect to *x*, we obtain

$$\cos(x+y)=-c_1\sin x+c_2\cos x.$$

Now let $x=0$ in the last two expressions. This gives us

$$\sin y=c_1 \text{ and } \cos y=c_2.$$

Therefore,

$$\sin(x+y) = \sin y \cos x + \cos y \sin x$$

and

$$\cos(x+y) = -\sin y \sin x + \cos y \cos x.$$

Hence result.

Note: We can use the above theorem to obtain the identities

$$\sin 2x = 2\sin x \cos x$$

and

$$\cos 2x = \cos^2 x - \sin^2 x = 2\cos^2 x - 1 = 1 - 2\sin^2 x$$

for all $x \in \mathbb{R}$.

6.3T8 Theorem:

There exists a real number π such that $2 < \pi < 4$, $\cos x > 0$ for all $x \in (0, \pi/2)$, cos is strictly decreasing on $[0, \pi/2]$, and $\cos(\pi/2) = 0$. Furthermore, sin is strictly increasing on $[0, \pi/2]$ and $\sin(\pi/2) = 1$.

Proof:

$$\frac{x^{2n+1}}{(2n+1)!} - \frac{x^{2n+3}}{(2n+3)!} = \frac{x^{2n+1}}{(2n+3)!}\left[2n+3-x^2\right]$$

$$\geq \frac{x^{2n+1}}{(2n+3)!}\left[5-x^2\right] > 0,$$

for all $x \in (0,2)$ and $n \in \mathbb{N}$. Therefore,

$$\sin x = \left(x - \frac{x^3}{3!}\right) + \left(\frac{x^5}{5!} - \frac{x^7}{7!}\right) + \left(\frac{x^9}{9!} - \frac{x^{11}}{11!}\right) + \ldots > 0$$

for all $x \in (0,2)$ and

$$\frac{d}{dx}(\cos x) = -\sin x < 0$$

for all $x \in (0,2)$. Hence, cos is strictly decreasing on $[0,2]$. If $x \in (0,2)$ and $n \in \mathbb{N}$, then

$$\frac{x^{2n}}{(2n)!} - \frac{x^{2n+2}}{(2n+2)!} = \frac{x^{2n}}{(2n+2)!}\left[2n+2-x^2\right] \geq \frac{x^{2n}}{(2n+2)!}\left[4-x^2\right] \geq 0.$$

Therefore,

$$\cos 1 = \left(1 - \frac{1}{2!}\right) + \left(\frac{1}{4!} - \frac{1}{6!}\right) + \left(\frac{1}{8!} - \frac{1}{10!}\right) + \ldots > \frac{1}{2} + \left[\left(\frac{1}{4!} - \frac{1}{6!}\right) + \left(\frac{1}{8!} - \frac{1}{10!}\right) + \ldots\right] > 0$$

and

$$\cos 2 = 1 - \frac{2^2}{2!} + \frac{2^4}{4!} - \left(\frac{2^6}{6!} - \frac{2^8}{8!}\right) - \left(\frac{2^{10}}{10!} - \frac{2^{12}}{12!}\right) - \ldots \leq 1 - \frac{2^2}{2!} + \frac{2^4}{4!} = -\frac{1}{3} < 0.$$

Since cos is continuous and strictly decreasing on $[1,2]$, there exist a unique $c \in (1,2)$ such that $\cos(c) = 0$. Define π to be the unique real number in the interval $(2,4)$ such that $\pi = 2c$. Then cos is strictly decreasing on $[0,\pi/2]$, $\cos x > 0$ for all $x \in (0,\pi/2)$, and $\cos(\pi/2) = 0$. sin is strictly increasing on $[0,\pi/2]$ because

$$\frac{d}{dx}(\sin x) = \cos x > 0 \text{ for all } x \in (0,\pi/2)$$

Furthermore, $\sin(\pi/2) = 1$ because $\sin^2(\pi/2) + \cos^2(\pi/2) = 1$, $\cos(\pi/2) = 0$, and sin is increasing on $[0,\pi/2]$.

6.3T9 Theorem:

The functions sin and cos are periodic with period 2π,

$$\max_{x\in\mathbb{R}} \sin x = 1, \quad \min_{x\in\mathbb{R}} \sin x = -1,$$

$$\max_{x\in\mathbb{R}} \cos x = 1, \text{ and } \min_{x\in\mathbb{R}} \cos x = -1.$$

Proof: Use (6.3T7b) to obtain

$$\sin(y + \pi/2) = \sin y \cos(\pi/2) + \cos y \sin(\pi/2) = \cos y \text{ for all } y \in \mathbb{R}.$$

Suppose $x \in [\pi/2, \pi]$. If $y = x - \pi/2$, then $y \in [0,\pi/2]$ and

$$\sin x = \sin(y + \pi/2) = \cos y.$$

Hence, sin is strictly decreasing on $[\pi/2,\pi]$.

$$\sin \pi = \cos(\pi/2) = 0.$$

Hence,

$$\sin x \geq 0 \text{ for all } x \in [0,\pi].$$

Furthermore,

$$\max_{x\in[0,\pi]} \sin x = 1 \text{ and } \min_{x\in[0,\pi]} \sin x = 0.$$

Now apply (6.3T7c) to obtain

$$\cos(y + \pi/2) = \cos y \cos(\pi/2) - \sin y \sin(\pi/2) = -\sin y \text{ for all } y \in \mathbb{R}.$$

It follows that cos is strictly decreasing on $[\pi/2,\pi]$, $\cos \pi = -\sin(\pi/2) = -1$,

$$\max_{x\in[0,\pi]} \cos x = 1, \text{ and } \min_{x\in[0,\pi]} \cos x = -1.$$

Furthermore, for all $x \in \mathbb{R}$

$$\sin(x+\pi) = \sin x \cdot (-1) + \cos x \cdot 0 = -\sin x$$

and

$$\cos(x+\pi) = \cos x \cdot (-1) - \sin x \cdot 0 = -\cos x$$

by (6.3T7b) and (6.3T7c). Therefore, the function sin increases from 0 to 1 on $[0, \pi/2]$, decreases from 1 to -1 on $[\pi, 3\pi/2]$, and increases from -1 to 0 on $[3\pi/2, 2\pi]$. The function cos decreases from 1 to -1 on $[0, \pi]$ and increases from -1 to 1 on $[\pi, 2\pi]$. For any $x \in \mathbb{R}$,

$$\sin(x+2\pi) = \sin x \cdot 1 + \cos x \cdot 0 = \sin x$$

and

$$\cos(x+2\pi) = \cos x \cdot 1 - \sin x \cdot 0 = \cos x.$$

Furthermore,

$$\sin(x+c) \not\equiv \sin x$$

and

$$\cos(x+c) \not\equiv \cos x,$$

if $0 < c < 2\pi$. Hence, both sin and cos are periodic with period 2π,

$$\max_{x\in\mathbb{R}} \sin x = 1, \quad \min_{x\in\mathbb{R}} \sin x = -1,$$

$$\max_{x\in\mathbb{R}} \cos x = 1, \text{ and } \min_{x\in\mathbb{R}} \cos x = -1.$$

6.3A3 Assignment:

(a) If $0 < x < \dfrac{\pi}{2}$, then show that $\dfrac{2}{\pi} < \dfrac{\sin x}{x} < 1$.

(b) Suppose $x \in \mathbb{R}$ and $n \in \mathbb{N}$. Use mathematical induction to prove that

$$|\sin(nx)| \le n|\sin x|.$$

(c) Give an example to show that the inequality in (b) does not hold if $n \notin \mathbb{Z}$.

6.3D5 Definition: Other Trigonometric Functions

Let

$$E = \mathbb{R} \setminus \{x | x = (2n+1)\pi/2,\ n \in \mathbb{Z}\}$$

and

$$F = \mathbb{R} \setminus \{x | x = n\pi,\ n \in \mathbb{Z}\}.$$

(a) The **tangent** function, denoted by tan, is defined on E and is given by

$$\tan(x) = \tan x = \frac{\sin x}{\cos x},\ x \in E.$$

(b) The **secant** function, denoted by sec, is defined on E and is given by:

$$\sec(x) = \sec x = \frac{1}{\cos x}, \ x \in E.$$

(c) The **cosecant** function, denoted by csc, is defined on F and is given by

$$\csc(x) = \csc x = \frac{1}{\sin x}, \ x \in F.$$

(d) The **cotangent** function, denoted by csc, is defined on F and is given by

$$\cot(x) = \cot x = \frac{\cos x}{\sin x}, \ x \in F.$$

6.3T10 Theorem:

(a) The functions tan and cot are periodic with period π with their ranges given by

$$\tan(E) = \cot(F) = \mathbb{R}.$$

(b) The functions sec and csc are periodic with period 2π with their ranges given by

$$\sec(E) = \csc(F) = (-\infty, -1] \cup [1, \infty).$$

Proof: Follows directly from (6.3D4) and (6.3T9).

6.3D6 Definition: Inverse Trigonometric Functions

The sine function is one-to-one, strictly increasing, and onto $[-1,1]$, if its domain is restricted to $[-\pi/2, \pi/2]$. The cosine function is one-to-one, strictly decreasing, and onto $[-1,1]$, if its domain is restricted to $[0, \pi]$. The tangent function is one-to-one, strictly increasing because

$$\frac{d}{dx}(\tan x) = \frac{\cos x \cdot \cos x - \sin x \cdot (-\sin x)}{\cos^2 x} = \sec^2 x,$$

and onto $\mathbb{R}$, if its domain is restricted to $(-\pi/2, \pi/2)$. Therefore, all three functions have inverses subject to these restricted domains.

(a) The **inverse sine** function, denoted by $\sin^{-1}$, is defined to be the inverse of the following function:

$$s(x) = \sin x, \ x \in [-\pi/2, \pi/2].$$

The domain of $\sin^{-1}$ is $[-1,1]$ and its range is $[-\pi/2, \pi/2]$. We can also define it as

$$y = \sin^{-1}(x) \Leftrightarrow x = \sin y \text{ for all } x \in [-1,1] \text{ and } y \in [-\pi/2, \pi/2].$$

(b) The **inverse cosine** function, denoted by $\cos^{-1}$, is defined to be the inverse of the following function:

$$c(x) = \cos x, \ x \in [0, \pi].$$

The domain of $\cos^{-1}$ is $[-1,1]$ and its range is $[0,\pi]$. We can also define it as

$$y=\cos^{-1}(x) \Leftrightarrow x=\cos y \text{ for all } x\in[-1,1] \text{ and } y\in[0,\pi].$$

(c) The **inverse tangent** function, denoted by $\tan^{-1}$, is defined to be the inverse of the following function:

$$t(x)=\tan x,\ x\in(-\pi/2,\pi/2).$$

The domain of $\tan^{-1}$ is $\mathbb{R}$ and its range is $(-\pi/2,\pi/2)$. We can also define it as

$$y=\tan^{-1}(x) \Leftrightarrow x=\tan y \text{ for all } x\in\mathbb{R} \text{ and } y\in(-\pi/2,\pi/2).$$

Note: The following results can easily be obtained using elementary methods:

(i) $\frac{d}{dx}(\sec x)=\sec x\tan x,\ \frac{d}{dx}(\csc x)=-\csc x\cot x,\ \frac{d}{dx}(\cot x)=-\csc^2 x.$

(ii) $\frac{d}{dx}(\sin^{-1} x)=\frac{1}{\sqrt{1-x^2}},\ \frac{d}{dx}(\cos^{-1} x)=\frac{-1}{\sqrt{1-x^2}},\ \frac{d}{dx}(\tan^{-1} x)=\frac{1}{1+x^2}.$

(iii) $\int \sin x dx=-\cos x+C,\ \int \cos x dx=\sin x+C.$

(iv) $\int \tan x dx=\log|\sec x|+C,\ \int \cot x dx=\log|\sin x|+C.$

(v) $\int \sec x dx=\log|\sec x+\tan x|+C,\ \int \csc x dx=-\log|\csc x+\cot x|+C.$

(vi) $\int \frac{1}{\sqrt{a^2-x^2}}dx=\sin^{-1}\left(\frac{x}{a}\right)+C,\ \int \frac{1}{a^2+x^2}dx=\frac{1}{a}\tan^{-1}\frac{x}{a}+C.$

6.3E5 Example: Find a Maclaurin series representation for $f(x)=\tan^{-1}(x)$.

The power series representation for the function

$$g(x)=\frac{1}{1+x^2},\ x\in\mathbb{R}$$

about 0 is given by

$$\sum_{n=0}^{\infty}(-x^2)^n=\sum_{n=0}^{\infty}(-1)^n x^{2n},\ |x|<1.$$

Therefore,

$$\int_0^x g(t)dt=\sum_{n=0}^{\infty}(-1)^n\int_0^x t^{2n}dt=\sum_{n=0}^{\infty}\frac{(-1)^n x^{2n+1}}{2n+1},\ |x|<1.$$

Hence,

$$\tan^{-1}(x)=f(x)-f(0)=\int_0^x g(t)dt=\sum_{n=0}^{\infty}\frac{(-1)^n x^{2n+1}}{2n+1},\ |x|<1.$$

Furthermore, since $\tan^{-1}$ is continuous and $\sum_{n=0}^{\infty}\frac{(-1)^n}{2n+1}$ converges,

$$\tan^{-1}(1)=\sum_{n=0}^{\infty}\frac{(-1)^n}{2n+1}$$

by (6.3T2). It is quite clear that $\tan^{-1}$ is an odd function. Therefore,

$$\begin{aligned}\tan^{-1}(-1)=-\tan^{-1}(1)&=\sum_{n=0}^{\infty}\frac{(-1)(-1)^n}{2n+1}\\&=\sum_{n=0}^{\infty}\frac{(-1)^n(-1)^{2n+1}}{2n+1}.\end{aligned}$$

Hence, the Maclaurin representation of *f* will be its power series representation about 0, and is given by

$$\sum_{n=0}^{\infty}\frac{(-1)^n}{2n+1}x^{2n+1},\ |x|\le 1.$$

This series converges absolutely on $(-1,1)$ and conditionally at the endpoints of the interval -1 and 1. Since $\cos(\pi/2)=0$,

$$2\sin^2(\pi/4)=2\cos^2(\pi/4)=1.$$

Hence,

$$\sin(\pi/4)=\cos(\pi/4)=1/\sqrt{2}\Rightarrow\tan(\pi/4)=1.$$

Therefore, we can represent π as the sum of a convergent alternating series:

$$\frac{\pi}{4}=\tan^{-1}(1)=\sum_{n=0}^{\infty}\frac{(-1)^n}{2n+1}$$

$$\Rightarrow\ \pi=\sum_{n=0}^{\infty}(-1)^n\frac{4}{2n+1}.$$

6.3D7 Definition: Hyperbolic Functions

(a) The **hyperbolic sine function**, denoted by sinh, is defined by

$$\sinh(x)=\sinh x=\frac{e^x-e^{-x}}{2},\ x\in\mathbb{R}.$$

(b) The **hyperbolic cosine** function, denoted by cosh, is defined by

$$\cosh(x)=\cosh x=\frac{e^x+e^{-x}}{2},\ x\in\mathbb{R}.$$

(c) The **hyperbolic tangent** function, denoted by tanh, is defined by

$$\tanh(x)=\tanh x=\frac{e^x-e^{-x}}{e^x+e^{-x}},\ x\in\mathbb{R}.$$

Note: We can also define the hyperbolic secant, cosecant, and cotangent in an analogous manner to the way that the secant, cosecant, and cotangent functions were defined. These functions will be denoted by sech, csch, coth, respectively. The following results can be obtained using elementary methods:

(i) $\frac{d}{dx}(\sinh x) = \cosh x,\ \frac{d}{dx}(\cosh x) = \sinh x,\ \frac{d}{dx}(\tanh x) = \operatorname{sech}^2 x.$

(ii) $\frac{d}{dx}(\operatorname{sech} x) = -\operatorname{sech} x \tanh x,\ \frac{d}{dx}(\operatorname{csch} x) = -\operatorname{csch} x \coth x,\ \frac{d}{dx}(\coth x) = -\operatorname{csch}^2 x.$

(iii) $\int \sinh x dx = \cosh x + C,\ \int \cosh x dx = \sinh x + C.$

(iv) $\int \tanh x dx = \log|\cosh x| + C,\ \int \coth x dx = \log|\sinh x| + C.$

(v) $\int \operatorname{sech} x dx = \tan^{-1}(e^x) + C,\ \int \operatorname{csch} x dx = \log\frac{e^x - 1}{e^x + 1} + C.$

6.3E5 Example: Let $f(x) = \cosh x,\ x \in \mathbb{R}$. Then

$$f'(x) = \sinh x = \frac{e^x - e^{-x}}{2} = 0 \Leftrightarrow x = 0$$

and

$$f'(x) < 0,\ x < 0 \text{ and } f'(x) > 0,\ x > 0.$$

Thus f is strictly decreasing on $(-\infty, 0]$, strictly increasing on $[0, \infty)$, and has a relative minimum at 0. If $ch : [0, \infty) \to \mathbb{R}$ is defined by

$$ch(x) = \cosh x,\ x \in [0, \infty),$$

Then ch is a strictly increasing function that is infinitely differentiable and its range is $[1, \infty)$, because $\cosh 0 = 1$ and $\lim_{x\to\infty} \cosh x = \infty$. Hence it has an inverse from $[1, \infty)$ onto $[0, \infty)$, denoted by $\cosh^{-1}$. This inverse is strictly increasing and (infinitely differentiable). Furthermore,

$$y = \cosh^{-1} x \Rightarrow x = \cosh y = \frac{e^y + e^{-y}}{2} \Rightarrow e^y + e^{-y} = 2x$$

$$\Rightarrow e^{2y} - 2xe^y + 1 = 0 \Rightarrow e^y = \frac{2x \pm \sqrt{4x^2 - 4}}{2} = x \pm \sqrt{x^2 - 1}.$$

Since $e^y > 1$ for $y > 0$, it follows that

$$e^y = x + \sqrt{x^2 - 1}.$$

Hence,

$$y = \log\left(x + \sqrt{x^2 - 1}\right),\ x \in [1, \infty).$$

6.3A4 Assignment: Show that the function sinh is strictly increasing on $\mathbb{R}$, $\lim_{x\to\infty}\sinh x = \infty$, and $\lim_{x\to-\infty}\sinh x = -\infty$. Deduce that $\sinh^{-1}$ exists and that the latter is a strictly increasing function from $\mathbb{R}$ onto $\mathbb{R}$ and find an expression for $\sinh^{-1} x$ as a log function.

6.3A5 Assignment: Establish the following identities using the results in this section:

(a) $\sinh(x \pm y) = \sinh x \cosh y \pm \cosh x \sinh y$

(b) $\cosh(x \pm y) = \cosh x \cosh y \pm \sinh x \sinh y$

(c) $\tanh(x \pm y) = \dfrac{\tanh x \pm \tanh y}{1 \pm \tanh x \tanh y}$.

Deduce that

(a) $\sinh 2x = 2\sinh x \cosh y$

(b) $\cosh 2x = \cosh^2 x + \sinh^2 x$

$$= 2\cosh^2 x - 1$$

$$= 1 + 2\sinh^2 x$$

(c) $\tanh 2x = \dfrac{2\tanh x}{1 + \tanh^2 x}$.